Aether:
Past, Present and Future
of the
Universe

Aether:
Past, Present and Future
of the
Universe

Bahram Esmailzadeh, M. Sc.

Library of Congress Control Number: 2012923597
ISBN: Hardcover   978-1-4797-6673-4
                Softcover   978-1-4797-6672-7
                Ebook   978-1-4797-6674-1

This book was printed in the United States of America.

For Information: Bahram Esmailzadeh
                  bahram965@gmail.com

To order additional copies of this book, contact:
Xlibris Corporation
1-888-795-4274
www.Xlibris.com
Orders@Xlibris.com
111883

This book is dedicated to,
    Every individual who has an open mind regarding new ideas.

Bahram Esmailzadeh

Acknowledgement

The author wishes to acknowledge that he is totally indebted to the Creator, who provided him with not only his existence, but also his sense of curiosity, his steadfast enthusiasm and interest, the required time as well as guidance to understand the lessons which were set before him.

The author also wishes to thank the Creator and all of the others, visible or not, for their direct guidance all through this particular research.

Table of Contents

Preface

The contents of this book and its companion book "'The Evolution of Spirits', What is the reason for being here, in this Creation?" are direct results of the author's personal thoughts and contemplations.

These books are intended to open up new doors in our journey as we try to understand the physical and the spiritual worlds, respectively. They introduce new ideas/theories which provide reasonable explanations for basic concepts in their respective areas.

Here, in this book, the author is only concentrating on the processes which have led to the formation of the physical universe and also how it is evolving. The details provided regarding various concepts clearly demonstrate the effectiveness of the theories presented. All of the theories are interdependent and supportive of each other.

Some of the chapters are written in a point by point style, as needed, to allow mandatory focus on the particular issues at hand. This method will also encourage the reader to stay focused.

For a quick overview it is recommended to read the "Introduction", sections on "The Evolution of the Universe", "What is Aether?" and "Predictions / Applications", as well as the "Conclusions".

In case more details are required, the reader is encouraged to read through the sections covering his/her specific subject of interest.

Introduction

In order to truly appreciate the greatness of this universe, one needs to get out of the populated areas, and on a clear and moonless night look at the sky. The sky will look like a carpet of lights, lights that are in fact stars. Stars that are visible to the naked eye actually make up a small portion of only one galaxy, namely the Milky Way Galaxy. It has been estimated that this universe contains over 10 billion such galaxies.

The universe exists due to a variety of reasons. We humans are just beginning to understand and appreciate a few of its laws, rules and regulations. Also, we are only aware of some of its contents and currently are guessing about the rest.

We have to admit that we are like children who have just started to understand what this universe is made of and what could be the purpose of its existence.

Even though we are learning quickly, our knowledge regarding the formation of this universe is still in its infancy. Therefore, we better have an open mind when we are faced with new but reasonable extensions to our current understanding of different aspects of the physical world.

Here, the author is presenting his own personal views and theories, as he is providing consistent explanations for a variety of phenomena that are commonly experienced and yet are not properly understood. Variety of phenomena such as time, light, space, gravity, electric field, magnetic field, black holes, the initial expansion process of the universe and even energy and matter, are shown to be interdependent.

The first chapter provides an overview of the evolutionary processes that this universe has experienced so far and what it will experience in the future. The second chapter provides detailed information about what aether is, what its properties are and how it relates to various physical phenomena.

The following chapters concentrate on specific phenomena, as they provide explanations on what each of those phenomena really are and how they relate to the other phenomena.

The last chapter provides a variety of predictions as well as applications that are made possible by the theories presented in this book. Some of the predictions are concerning the current unresolved issues such as the widening of the planetary orbital paths, while the others are in regards to events that have already taken place or will take place later on in the future.

One thing that all of the sections in this book have in common is the existence of aether, the very same aether that was believed to exist by the scientists in the nineteenth century and the early twentieth century.

As it is demonstrated throughout this book, once the existence of aether is accepted and its effects are taken into account, many unresolved issues regarding various basic phenomena will be explained and readily understood.

This book is written for the general public. However, possessing some general background knowledge of the topics covered will be helpful in appreciating the simplicity of the theories introduced, and the straightforward explanations that are presented.

The Evolution of the Universe

The Evolution of the Universe

The evolution of the universe should be studied in several distinct periods. The beginning/formation era, current situation as well as the future/destiny of the universe should be considered separately. But, first one needs to define what 'space' is.

Space

Space owes its existence to aether. Aether, as it is described in great detail in another section, "What is Aether?" is an elastic, compressible, pressurized fluid that literally occupies the whole universe. The relationship between space and aether is just like that of an ocean and water. The existence of water literally gives meaning to an ocean. If all of the water that is inside an ocean is taken out, what is left can no longer be called an ocean. The very same is also true about aether and space. For space to exist, it must be occupied by aether.

The physical space is not limited or restricted to the three spatial dimensions that jointly define what is commonly referred to as the physical universe. Space also includes three hidden dimensions, which are accessible under certain conditions. The hidden dimensions form a universe that is accompanying this universe.

Note that, even though the accompanying universe is a parallel universe to this one, it is not hosting a duplicate set of everything that exists in this universe. For one, currently the density/pressure of aether in the accompanying universe is much lower than the density/pressure of aether in this universe.

The two universes are connected together through infinite number of matter particles which act as drain holes for aether to escape from this universe and enter the accompanying universe. The two universes are also connected through black holes which simply are large aggregates of matter particles that have managed to unite and form a bigger opening for aether to flow through.

The accompanying universe has its own set of duties to perform. It is in fact a complementary part of this universe. According to the information presented here in this book, the accompanying universe is needed for this universe to function properly.

The Beginning

In the beginning, aether was compressed into a very small volume. In other words, space was quite limited in its overall size. Aether's density/pressure was enormous. One could say that it was actually at its critical point.

Somehow a ripple effect was introduced into the aether medium, just like when a pebble is dropped into the middle of a pond. The ripples caused the over excitation of the aether medium which started to expand (not explode, but expand) in all spatial directions.

This marked the beginning of the rapid expansion era.

When the expansion started, the speed of the propagation of any type of phase vibrations, such as the ripples in that medium was at an all time low, due to high aether density. This meant that, in the beginning, the ripples spread more by being carried by the expanding aether medium rather than by travelling in it.

In other words, the ripples expanded both with as well as in the aether medium, as the aether medium (space) itself was expanding.

The rate at which 'time' is experienced (by any entity) is dependent on the speed (of that entity) relative to the local aether medium. Please refer to section "What is the Essence of Time?" for greater details. As this relative speed approaches that of phase vibrations in the medium of aether, time is correspondingly experienced at a slower pace.

The slow speed of ripples (phase vibrations) spreading in the aether medium also meant that "time" was experienced at a much, much slower pace back then as compared to the present time. In other words,

"The duration of the first 'second' of the aether expansion process was equivalent to what may be millions or even billions of years long, based on today's time scales.

The expansion process of aether, due to not facing any kind of resistive force, such as gravity, caused this universe to go through an enormous growth burst. In other words,

"The very lack of the force of gravity, allowed the unrestricted expansion of the aether medium (space)."

This growth burst, in a relatively short period of time (based on then-current time scale), spread the aether and hence expanded the space from a very small region to a volume that was literally equivalent to billions of present-time light years across.

As aether expanded, the ripples spread and stabilized in their patterns. These ripples were in fact phase vibrations in that medium. Phase vibrations in aether are just like regular sine waves. They do have a positive half and a negative half, during every complete cycle.

Under certain conditions, phase vibrations in the aether medium can be encouraged to form full wave as well as half wave spikes and so lead to the formation of concentrated/condensed states of aether, in that medium. Simple presentations for both full wave as well as half wave spikes are shown below.

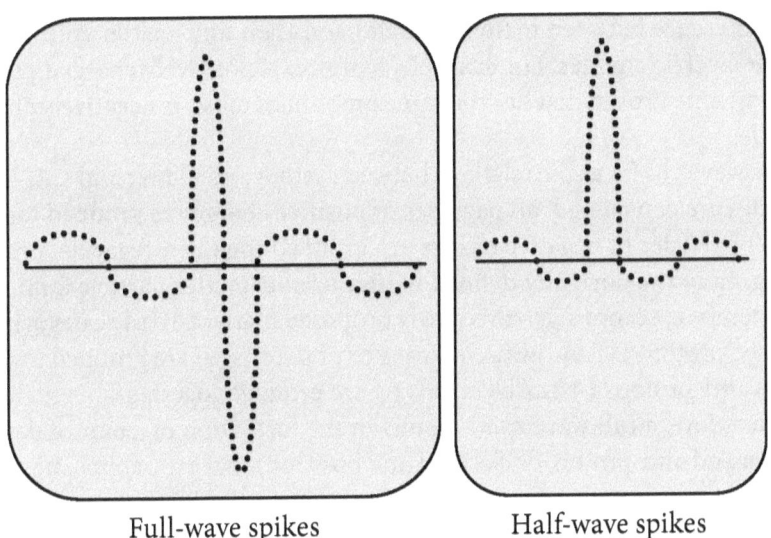

Full-wave spikes   Half-wave spikes

At a certain stage of the expansion process, the ripples in aether were provided with the needed conditions and so automatically and simultaneously started forming local eddies everywhere. These local eddies eventually led to the formation of spikes.

When spiked (condensed), the positive halves lead to the formation of matter particles that are of positive (+) charge and the negative halves lead to the formation of matter particles that are of negative (-) charge. The matter and anti-matter particles that exist in this universe are literally condensed forms of aether that are floating in the aether medium.

The formation of matter and anti-matter particles in aether as well as the timing of their appearance can be likened to the formation of icicles from individual molecules of water, as the temperature of the medium of water is lowered. As a certain temperature is reached, the entire volume/body of water simultaneously starts to crystallize.

This is basically what happened within the medium of aether. As its density/ pressure was decreased due to its expansion process, it literally provided the needed environment for such stable eddies to form and eventually develop into matter and anti-matter particles.

Note that, even though ice is less dense as compared to water, condensed states of aether corresponding to matter and anti-matter particles are much denser than aether in its normal fluid state.

According to the currently accepted definitions in particle physics, the only difference between matter particles and their anti-matter counterparts is their electric charges. For example, a proton is positively charged particle, while an anti-proton that has the same amount of mass, is negatively charged particle.

However, as far as the relations between aether and matter and anti-matter particles are considered, all particles of positive charge are grouped together and all particles of negative charge are grouped together, regardless of their belonging to the normally defined matter or anti-matter particle families.

Therefore, according to this newly proposed matter particle categorization strategy, protons (+) and anti-electrons or positrons (+) are grouped together, just as anti-protons (-) and electrons (-) are grouped together.

Therefore, a full wave spike results in the formation of a pair of particles (proton and anti-proton, or electron and positron) that are of opposite electric charges. Correspondingly, a half wave spike results in the formation of a single particle that is either positive or negative in charge, depending on the half wave being positive or negative.

Matter and anti-matter particles are basically the byproducts of the full wave and half wave spikes that were formed in aether. As a result, when the conditions were right, the whole medium of aether was literally suddenly saturated with matter and anti-matter particles.

Note that, the types of particles formed during these processes solely depended on the sizes and types of spikes that were formed. For instance, the more pronounced full wave spikes gave birth to proton and anti-proton pairs, while the weaker ones could only promote the formation of electron and positron pairs, and so on. In other words,

**"Density spikes in aether manifest themselves as
matter and/or anti-matter particles."**

Since the phase vibrations were distributed uniformly in the
aether medium, the same circumstance was provided throughout
its volume. Therefore, the phase vibrations present in aether resulted
in the formation of matter and anti-matter particles everywhere,
at the same time. In other words, one could say that,

**"The whole universe literally bloomed across its volume with
matter and anti-matter particles, at about the same time."**

As it is explained in the section "What is Gravity?", matter and anti-
matter particles act like tiny holes in the fabric of space (not in the aether
medium) allowing the aether that is at higher pressure in this universe, to
escape into the accompanying universe where its pressure is much lower.

One can visualize each of the spikes in the aether, forming tiny tubes
(tunnels) between this universe and its accompanying universe. The very
formation of such tubes automatically allows the flow of aether from this
universe to the other one, simply due to the pressure difference that exists
between the two universes.

Matter and anti-matter particles basically acted (and are still acting) as
pressure relief valves. Right after their formation, by allowing the rapid leakage
of aether from this universe (due to their enormous population throughout this
universe), matter and anti-matter particles caused three very important effects:

1- They encouraged a very quick drop in the density/pressure of aether
in this universe.

2- The instinctive flow of aether towards matter and anti-matter
particles exerted a mutual drag force on the neighboring particles
and eventually on all existing particles in this universe.
This very action of aether, namely its flow towards particles,
literally gave meaning to what is now known as the force of gravity.
In other words,

**"The very formation of matter and anti-matter particles
marked the Birth of the Force of Gravity."**

For detailed information on gravity, please refer to the section
"What is gravity?"

3- **The very slower motion of the matter and anti-matter particles relative to the local aether gave meaning to "TIME".** As it is explained in great details in a separate section "What is the Essence of Time?", time is only experienced by any and all entities that move (relative to aether) at a speed that is slower than the speed of the phase vibrations in the local aether. In other words,

> **"Time started as the first ripples were formed and started to propagate in the aether medium."**

However,

> **"The meaningful startup point for experiencing time was when matter and anti-matter particles were formed, since as they fell behind as compared to the ripples that had formed them, they were the very first entities that experienced the passage of time."**

With the formation of matter and anti-matter particles which encouraged the leakage of aether to the other universe, as well as the expansion of the medium itself, the pressure of the aether medium was lowered drastically.

Shortly after coming into existence, most of the matter and anti-matter particles annihilated each other. Their annihilation process which was in fact cancelling of each others' wave functions in aether literally spread their aether contents and saturated the aether medium with high frequency, but much lower amplitude, phase vibrations. A simple presentation of such a process is shown below.

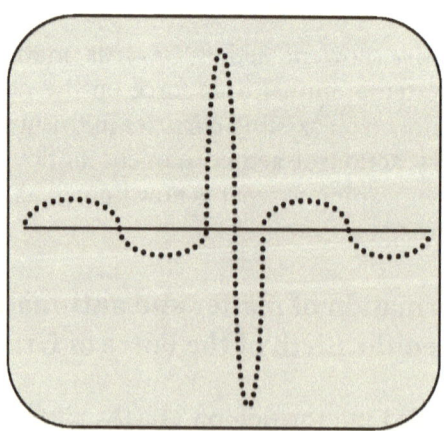

Before Annihilation

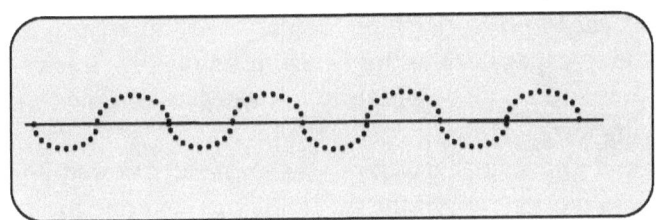

After Annihilation

Note that, the phase vibrations generated due to the annihilation of the matter and anti-matter particles were further spread and flattened by the expansion of the aether medium.

Over time, these phase vibrations spread everywhere and literally became a uniform background noise in the medium of aether. At the present time, they are detected as **cosmic background radiation** that exists literally throughout the whole universe.

The neutral particles, namely the neutrons, where formed from the joining of electrons and protons (or anti-protons and positrons), as they attracted each other due to having unlike charges.

Note that, the artificial generation of massive particles in various particle accelerators is only indicative of the formation of ever more pronounced (higher amplitude) full wave as well as half wave spikes in the phase vibrations that exist in the aether medium. However, by no means, they are representatives of the previous generations of particles that supposedly by splitting have led to the formation of the present-day particles.

The formation of heavier particles in particles accelerators can be simply looked at as if someone is literally turning up the "volume" on a stereo system or as if more speakers are used in sync with each other to generate a specified vibration but at higher amplitudes. The newly formed particles (spiked phase vibrations) have literally nothing to do with the reality of what has naturally existed in this universe. In other words,

"The massive particles generated in particle accelerators do not represent what naturally have existed or will exist in this universe."

Matter and anti-matter particles have always acted (and are still acting) as pressure relief valves. By their very nature, matter and anti-matter particles caused the density/pressure of the aether medium (which was already dropping due to expansion) to drop even more rapidly. This rapid drop in the aether pressure led to a rapid drop in its expansion rate, since pressure was its driving force.

Matter and anti-matter particles also drastically slowed down in their own spreading motion within the medium of aether, due to the enormous force of gravity that they were exerting on each other. Therefore, one could say that,

The formation of matter and anti-matter particles, literally acted like an effective braking system because,

- **It drastically slowed down the spreading of the remaining particles, due to the force of gravity,**

- **It drastically slowed down the expansion of the aether medium as a whole, by reducing its internal pressure.**

As aether continued to expand, after most matter and anti-matter particles had already annihilated each other, it became a calm and hospitable environment for the surviving matter particles. These remaining matter particles continued with their expansion process, but at a much, much slower pace. In other words,

This period marked the end of the rapid expansion era.

The very process of expansion that the infant universe experienced was exactly like what is experienced by the critical mass inside a nuclear bomb, as it explodes. As an external neutron source is introduced into its center, the critical mass suddenly gets the urge to start a progressively more violent nuclear chain reaction. The material forming the critical mass experiences a sudden expansion that is literally near the speed of light. In the mean time, the shock wave which starts at the center also starts to propagate through the expanding material.

As it is viewed from distance, a nuclear explosion goes through a very sudden but temporary growth burst (at nearly the speed of light) which is followed by a continuing but very gradual expansion of the byproducts of the explosive material.

This is exactly what took place in this universe, during its infancy period.

However, since "time" was experienced at a much, much slower pace as compared to the rate it is experienced at the present,

The first moments of expansion took equivalent of possibly millions, if not billions, of years based on today's time scales.

Matter particles that survived the annihilation period had lost most of their initial outwards momentums, due to the tremendous force of gravity that existed while matter and anti-matter particles were still in abundance. Since their speeds were much lower than the speed of the phase vibrations in that medium, the matter particles literally fell behind from the phase vibrations (ripples) that had formed them.

Therefore, as they were still being dragged by the expansion of the aether medium (space), they also had their own initial momentums/ movements within the aether medium. In other words, they started their semi-independent wandering state in aether.

Each and every matter particle was moving at a different speed relative to its local aether medium, as well as relative to each other. However, their general direction of motion was still away from the center, where the aether expansion had started from.

The overall expansion rate of the matter particles was controlled by the following opposing factors:

1- Their initial outwards momentums that were still moving them farther apart from each other.

2- Their mutual force of gravity that was counteracting their initial outwards momentums.

The effectiveness of the force of gravity between any two objects depends on two factors:

- The force of gravity depends on the density/pressure of the local aether, since denser medium develops a more effective dragging force, just as more viscous fluids do as compared to less viscous fluids.

  The aether's density/pressure is inversely proportional to the cube of any changes that occurs in its medium's diameter.

- The force of gravity between two objects also depends on the inverse of the square of their respective distances, due to the equation defining the surface area of a sphere. (4π)

The variations in the surface area of a sphere, through which aether approaches a given particle, are directly dependent on the square of its radius.

The above mentioned effect due to changes in aether's density as a result of expansion was cancelled by the corresponding reduction in the internal pressure of the aether medium which was the driving force for its expansion.

In other words, as aether's pressure, the driving force for its expansion was reduced, so was aether's density which was directly affecting the induced drag force that was the force of gravity. In fact, as aether medium was expanding, its density and its pressure were literally changing by exactly the same amount. Therefore, the overall expansion rate of the remaining particles was literally affected only by their distances, due to their mutual force of gravity.

However, the distances between the remaining matter particles were increasing both due to particles' motions (spreading) in aether and due to the expansion of the aether medium (space) as a whole. The combination of these two effects leads to an apparently weaker force of gravity than expected over long distances. This is due to the fact that, longer distances are communicated over longer periods of time which include older times, when aether was also denser. In other words,

On large intergalactic scales, the force of gravity changes slower than it is predicted by the inverse squared law.

$$F \approx Gm_1m_2 \,/\, x^{1.999}$$

Likewise, over short distances the force of gravity changes faster than it is predicted by the inverse squared law. This is due to the gradient that exists in the density profile of the aether as it approaches a large aggregate of matter particles, such as a star or basically within a given galaxy. For more details please refer to the section on "What is Gravity?" In other words,

On smaller scales (within galaxies), the force of gravity changes faster than it is predicted by the inverse squared law.

$$F \approx Gm_1m_2 \,/\, x^{2.001}$$

The density/pressure of the aether medium is still continually decreasing due to two reasons:

1- The expansion of the overall aether medium, due to its internal pressure.

2- The leakage of aether through the matter particles which act as drain holes for aether to flow into the accompanying universe.

The expansion process of the aether medium (space) will continue for a long time into the future, but at ever slower paces. This is due to the fact that, aether's expansion is purely pressure driven. Therefore, as its pressure is reduced, its expansion rate will also be reduced.

As the density/pressure of the aether medium is gradually reduced, the speed at which phase vibrations such as electromagnetic waves (light and so on) travel through aether is increased. In other words,

"The speed of light is gradually increasing."

According to the information provided in the section on "What is the Essence of Time?", the rate at which time is experienced by any object (including living beings) is directly dependent on that object's speed relative to its local aether medium. As this relative speed approaches the speed of the phase vibrations in aether, time is experienced at a slower pace.

Therefore, as the speed of the phase vibrations in the aether medium is increasing (due to decreasing density of aether medium), but the speed of the matter particles in aether is not changing by much (due to their mutual gravitational forces), the rate at which the passage of time is experienced is actually gradually speeding up. In other words,

"Time is speeding up."

The increase in the rate at which time is experienced by the contents of this universe and the increase in the speed of light over time, can give (and it is giving) the impression that the expansion rate of the universe, as a whole, is speeding up. In other words,

The universe, as a whole, is continuing to slow down in its expansion rate, not just due to having enough matter to slow

it down, but more effectively so due to losing internal aether pressure which is the driving force for its expansion.

Over billions of years, following the rapid expansion ear, due to the gravitational forces that the remaining matter particles were exerting on each other, they formed local groups and eventually formed huge, local giant cloud-like gatherings, throughout the aether medium. Since individual particles that were present in each of these cloud-like gatherings were moving at different speeds, they automatically caused their collectives to rotate. This joining process is shown below for the simplest case that involves a proton and a neutron with different speeds relative to each other.

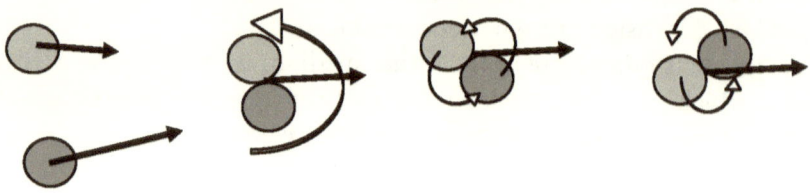

The only major force governing the motion of these particles was the force of gravity. Their electrical force fields came into play only when they were both charged particles, or charged particles joined a group that already had a charge, as a whole.

Over time, the force of gravity also encouraged the formations of numerous even more localized concentrations of matter particles within each of these giant cloud-like gatherings. As these localized concentrations of matter particles became denser and denser, they eventually led to the formation of stars.

Again, the differences that existed in the speeds of individual particles, that joined together, caused their collectives, namely the stars to rotate.

Due to the nuclear reactions and hence the super high temperatures that exist at their cores, stars are in fact not only the generators of enormous amounts of heat, but also the generators of all elements which exist in nature. By combining the nuclei of lighter elements such as Hydrogen, Helium and so on together, step by step, stars have managed and are still managing to produce each and every nucleus of all elements in existence in nature.

Because of the ongoing nuclear reactions within stars, their surface temperatures are always at several thousands of degrees Kelvin and their inner core temperatures are in the range of millions of degrees Kelvin.

In the meantime, due to the enormous size of the nuclear reactions which are continuously in progress in all stars, they induce strong magnetic and electric fields around themselves.

Huge explosions that occur relatively close to the surfaces of stars, due to violent electric and magnetic field interactions, result in all kinds of particles and elemental byproducts to be literally thrown into the surrounding space. These particles depart from their parent stars in the form of particle storms (solar storms).

Due to the force of gravity, these particles slow down, as they get farther and farther away from their respective stars. They start to follow an orbital path around the star that has manufactured them. Over time, as they exert their force of gravity towards each other, these particles form localized groups of their own, while orbiting their star.

The formation of the planets has been due to such gatherings of more and more particles floating in space, as they are concentrated into localized masses, over billions of years.

Notes,

1- Due to the differences that exist between the speeds of these freely floating particles, as they get together and form a union such as a planet, their collectives have a rotation associated with them.

2- As time goes on, due to absorbing all of the particles that are literally thrown into their path, every planet is in fact continuously gaining weight.

3- The lighter elements and particles in general are thrown farther as compared to the heavier ones. This is clearly manifested in the overall density profiles of the planets that exist in our solar system.

The overall densities of the three inner planets are around 5.3 g/cm3, while the overall densities of the outer planet average about 1.5 g/cm3 and that of Mars is at 3.93 g/cm3.

Over time, billions and billions of solar systems have formed through these processes. Later on, as many, many stars and solar systems became more organized, they formed the galaxies.

In other words, galaxies are the final results of the evolutionary processes that the local cloud-like gatherings have gone through, over billions of years. They started from a totally chaotic state and due to the gravitational interactions between their constituents, managed to develop into well behaved systems of celestial bodies which continue to hold on to each other with the help of their force of gravity.

Stars have been the generators of heat as well as the nuclei of all of the natural elements. Planets, on the other hand, due to their surface temperatures being much lower than those of the stars, over time, have performed their own unique tasks. Lower temperatures on planets have allowed the formation of atoms, molecules which in turn led to the formation of gases, liquids, and in some cases even solids, with different compositions.

Accumulation of gases led to the formation of gaseous layers around most planets and eventually formed their atmospheres which gradually stabilized in their compositions. During the very same period, the composition of the surface materials went through their own specific transformations.

Surface and atmospheric conditions on some of the planets led to the formation of water and other chemical compounds, such as the amino acids. Such chemical compounds by being in certain environmental conditions, and with the direct cooperation of the contents of the spiritual space, have led to the formation and development of various types of plants and animals.

Current Situation

Currently, the contents of this universe, on different scales, are fairly stabilized and abide by certain rules and laws, which are the same everywhere. For example,

- All physical entities experience the passage of time. Even though, due to either moving at an incredibly high rate of speed, or being in strong gravitational, magnetic or electric fields, they can experience the passage of time at a slower pace, but they all experience the passage of time in the positive direction (always towards the future).

- All physical and chemical laws are the same across this universe.

- The atoms belonging to any element have the very same structure and exhibit the very same physical and chemical properties, regardless of where in this vast universe they were formed.

Destiny of the Universe

Over time, massive stars will go supernova, as many already have. During this process, while their outer layers get literally thrown into the surrounding space, their inner parts implode and give birth to black holes. Black holes can only grow in size. They will keep on devouring any and all of the planets and stars that get close enough and get trapped in their one-way aether flow.

Black holes will eventually devour all of the matter particles in their vicinity. They will also form giant black holes, as they join with each other. Eventually, black holes will possibly be amongst the final pieces left in each galaxy.

Note that, the effectiveness of any black hole is directly dependent on the local aether pressure difference between this universe and the accompanying universe.

As time goes on and the aether expands (and also leaks to the accompanying universe) its pressure in this universe is continually dropping, while the pressure of the aether medium in the accompanying universe is rising.

Note that, lower aether density/pressure in this universe automatically implies higher phase vibration speeds (faster light speeds) in this universe.

In the future, the pressure difference between the aether medium in this universe and the accompanying universe will approach a critical value. At that point, the pressure difference will no longer be able to encourage the flow of aether towards even the largest possible black holes to reach the speed of light (the speed of phase vibrations) in the local aether medium. This critical aether pressure difference can be referred to as,

"Esmailzadeh Critical Aether Pressure Difference"

As this critical pressure difference is approached, all of the black holes, starting with the smallest ones, will lose their status as being black holes, since light among other electromagnetic waves will be able to travel against the flow of aether and escape. In other words,

"Eventually, all black holes will become visible."

As the aether's density is decreasing over time, due to its expansion as well as leakage (to the accompanying universe), it is becoming less effective in dragging objects that happen to be in its path. Therefore, stars are gradually, exerting less of a gravitational force on their respective planets. This is in turn leading to the widening of the orbital path of planets around stars.

One can say that, stars are gradually losing their grip on their planets and consequently

"The planetary orbits are gradually becoming wider."

This side effect of decrease in aether's density has already been detected (in regards to the planet earth). According to the collected data, the orbital path of the planet earth is widening by about 7 meters (about 23 feet) per century.

Of course, a small portion of the observed amount is justifiably due to the sun losing mass as it is continuously transforming some matter into energy and also as it is literally throwing some matter into space as solar storms.

Note that, the widening of the orbital path of planets around their respective stars will lead to eventual and definite **GLOBAL COOLING** on all planets. This global cooling, which will actually take place on all planets existing in this universe, can be referred to as,

"Scharback Universal Planetary Cooling"

As the pressure/density of aether in this universe is reduced further and the force of gravity becomes even weaker, stars will no longer be able to hold on to their respective planets. Stars also will no longer be motivated to stay in their own orbits around the centers of their respective galaxies, either. Therefore, solar systems and galaxies will lose their structural integrities and all planets and stars will relatively freely float in space. In other words,

"All of the celestial bodies will become independent wanderers in this universe."

As the force of gravity becomes weaker, the outer layers of stars will no longer weigh as much. Therefore, they will not exert enough pressure on the inner layers to promote the ongoing nuclear fusion reactions. At this point, stars will literally turn off and start to cool down. They would basically become cold gas giants.

As all of the stars, one by one, turn off, **the lights in the night sky will gradually become less in numbers and eventually all will disappear, forever.** Once all of the stars are turned off, the whole universe will start its eternal dark era. In other words,

"The universe will become pitch black"

Also,

"As stars such as our sun turn off, their respective planets like our earth (if they are still in orbit) will start experiencing a night that is eternally long."

The pressure of the aether in this universe will keep on dropping until it equalizes with the pressure of the aether medium that is in the accompanying universe.

At this point in time, there will be no motivation (pressure difference) for aether to flow towards matter particles or black holes, and enter the accompanying universe. Therefore, as the flow of aether towards matter particles, and even black hole, comes to a complete halt, it automatically implies that, the force of gravity will cease to exist. In other words,

"The force of gravity will gradually, literally fade away."

At that point in time, if the universe has any momentum left in its expansion, it will keep on expanding forever, since due to the lack of gravitational forces, the universe would literally lack its braking instincts. This era in the existence of this universe can be referred to as,

"Esmailzadeh Zero Gravity Era"

From this point on, not only none of the existing matter particles will exert any gravitational forces on each other, but also all of the matter particles that were trapped inside black holes will be able to freely literally float away, either towards this universe or the accompanying universe.

This final stage will eventually lead to a homogenous distribution of matter particles that will be freely floating in total harmony and peace, without exerting any kind of force towards each other.

Note that, even though there would not be a net flow of aether between this universe and its accompanying universe, yet due to the remaining pressure in the aether medium, it will keep on expanding literally forever.

Ultimately, as its density/pressure approaches zero, the overall size of the aether medium (space) will approach infinity.

Also, as the aether's density is approaching zero, the speed of the phase vibrations in that medium will correspondingly approach infinity. In other words,

"The speed of light and other phase vibrations will approach infinity."

And, as the speed of phase vibrations in the medium of aether approaches infinity, the rate at which time is experienced in this universe will also approach infinity. In other words,

"Gradually the duration of minutes, hours, days, years, even millions of years will become equivalent to the duration of one 'second', as it is experienced today."

What is Aether?

What is Aether?

In 1865, Mr. Maxwell proposed his theory on electromagnetism. Using his formula, he quite accurately calculated the speed of light in a vacuum. Mr. Maxwell had based his theory on the existence of some kind of medium through which light and other electromagnetic waves were travelling.

This medium at that time was referred to as Aether. What aether was made of, or what kind of properties it had was not known. It was necessary to be there as a medium for electromagnetic waves to travel in. The aether to electromagnetic waves was thought to be just as air is to sound waves.

Over time the scientists became curious about aether and tried to understand its behavior. They thought if aether was stationary in space and all waves and even planets were passing through it, then one should be able to detect the drift effect on the surface of the earth. Knowing that earth is orbiting the sun at about 107,000 Kilometers per hour, various attempts were made to detect this drift velocity.

The most promising of all the proposed experiments were designed and performed by Mr. Michelson and Mr. Morley in 1887. Again, these two gentlemen had assumed that aether was stationary in space and earth was moving through it. They had designed their apparatus with sufficient accuracy to detect such a minute velocity variation caused by the earth's motion around the sun as compared to the speed of light in a vacuum.

Their experiment was based on the comparison of wave patterns created by the same monochromatic light passing through two paths, one in the direction of the earth's motion around the sun and the other at 90 degrees to the said direction. By reflecting these two light branches back and superimposing them, Mr. Michelson and Mr. Morley were hoping to see the formation of some sort of interference between the two.

They conducted their experiment at various locations on the surface of the earth. They also conducted their experiment up at high altitudes, with the help of a balloon. However, they did not observe any interference pattern in any of their attempts.

Of course, some scientists suggested that the planet earth could be carrying a thin layer of this aether medium with it, as it is going around the sun. This could explain why no effect was detected.

In 1904 two scientists, Mr. Fitzgerald and Mr. Lorentz, independently, provided reasoning for detecting no trace of any drift effect in all of the experiments. Mathematically, they proved that the length of the experimental apparatus's arm which was in the direction of earth's motion was affected by the motion and therefore had resulted in annulling the intended effect of the aether drift.

Mr. Lorentz, in 1904, proposed a theory based on which the effect of motion on the apparatus was clearly explained. This theory was the first theory ever to talk of effects observed due to motion at relatively high velocities.

Particularly, the effects such as the shortening of the length of the object in the direction of motion, as well as the slowing down of the passage of time as one approaches the speed of light in a vacuum, were first proved by Mr. Lorentz' theory and his transformation equations. These transformation equations are also known as Mr. Lorentz' Relativity Theory or "Principles of Relativity".

Again, it should be noted that, Mr. Lorentz and Mr. Fitzgerald had proposed their theories without rejecting the existence of aether, which was commonly accepted by the then-current scientific community.

In 1905, Mr. Einstein used Mr. Lorentz' "Principles of Relativity" and added the following two assumptions:

1- Mr. Einstein rejected the existence of the aether altogether and claimed that light is made of particles, namely photons.

2- Mr. Einstein claimed that the speed of light in a vacuum is the maximum possible speed, literally for anything and everything in this universe, and it is a constant that is independent of the observer's state of motion relative to the source.

Mr. Einstein did not provide any proof, as he added these extra assumptions to the already existing Theory of Relativity that was proposed by Mr. Lorentz. But, he did give the revised version of the Principles of Relativity a new name, namely "Special Theory of Relativity".

Note that:

1- The experiments performed by Mr. Michelson and Mr. Morley, by no means had indicated that there was no aether in space, or that the speed of light was the maximum speed possible.

A revised version of the experiment performed by Mr. Michelson and Mr. Morley is proposed in a separate section of this book.

2- Since Mr. Zwicky, a Swiss astrophysicist, published his report on gravitational effects which were due to some kind of invisible mass (matter) in 1933, two new terms have been added to astrophysicists and physicists' vocabulary, namely dark matter and dark energy. Dark matter apparently does not directly interact with known forms of matter. Yet, its gravitational effects on known matter have been observed since Mr. Zwicky's time.

The relative amounts of dark matter, dark energy and regular matter/energy in this universe are estimated to be:

73%, dark energy,
23%, dark matter, and
4%, all known forms of energy and all known forms of visible matter

It is not really important to refer to aether with its original name, as aether. **The main point is that, the scientific community has to accept that aether exists**, just as the 19th century scientists believed it did, and had formulated their theories based on its existence. Maxwell's electromagnetism (1865) and Lorentz's principles of relativity (1904) (not Einstein's special relativity (1905)), were among such theories, theories that are still valid to this day and are in widespread use.

Understanding Aether

In this book, a variety of theories are introduced that are based on the existence of aether in this universe. These theories explain the very essence of various phenomena such as "light", "gravity", "magnetic field", "electric field", "black holes", "sudden expansion of the universe" and even "time", "space", "energy" and "matter".

These phenomena not only rely on the existence of aether. They also rely on each other and affect each other in measureable ways. The following are brief descriptions of how aether relates to these phenomena.

- **Space,** is filled with aether and is literally dependent on its presence. In other words,

"Space without aether is literally meaningless."

- **Time** is experienced (by an object or a living being) only if its motion relative to the local aether is at a speed that is less than the speed of phase vibrations in that medium.

  As its speed approaches the speed of light (the speed of phase vibrations in the local aether) the pace at which the passage of time is experienced by the object or living being also slows down. And, if its speed becomes equivalent to that of phase vibrations in the local aether medium, that particular object or living being will no longer experience the passage of time.

  Note that, the very existence of aether and phase vibrations in it are the prerequisites for time to exist. Therefore, 'time' started its existence as the first phase vibrations were formed in aether.

- **Energy** is simply the variety of motions, such as translational or vibrational motions, that exist in the medium of aether.

  Energy is manifested in many different forms and yet, when

considered on a microscopic level, they are all one form of motion, or another, of aether.

- **Dark Matter** is one of the most common forms of aether in this universe.

- **Matter** is a condensed form of aether, as it literally spikes in concentration. The positively and negatively charged matter (as well as anti-matter) particles are different portions of the full-wave phase vibrations in aether, as they spike.

  One can visualize each of the spikes in the aether, forming tiny tubes (tunnels) between this universe and its accompanying universe. The very formation of such tubes automatically allows the flow of aether from this universe to the other one, simply due to the pressure difference that exists between aether that is in the two universes.

- **The force of gravity** is due to the drag force induced by the linear flow of aether at any given location in space.

  As aether flows towards subatomic particles (which act as drain holes), it literally drags anything and everything that happens to be in its path. The influence of this kind of motion created in aether is long-range and reaches the end of space.

  In this case, the flow of aether is a one way trip towards subatomic particles. As aether reaches the subatomic particle, it literally goes through it and enters the accompanying universe which is also made of 3 spatial dimensions.

- **Magnetic field** is a locally induced round trip flow, in aether, a flow that actually forms a complete loop.

  In this case, the flow of aether is continuously within this universe and it is only a local <u>short to mid-range</u> disturbance created in the aether medium.

- **Electric field** is a locally induced round trip flow, in aether, a flow that actually forms a complete loop.

  In this case, the flow of aether is continuously within this universe and it is only a local <u>short-range</u> disturbance created in the aether medium.

- **Black holes** are enormous drain holes made of countless number of matter particles that have literally joined together and are acting as a unified gate through which aether flows to some other dimension, namely the accompanying universe.

As aether is flowing towards black holes (which act as huge drain holes), it literally drags anything and everything that happens to be in its path.

In this case, the flow of aether is a one way trip towards the black hole. As aether approaches a black hole its flow speed increases and at some point (distance from the black hole) it reaches the speed of the phase vibrations in aether. That particular distance to the black hole is referred to as the event horizon of the black hole. As aether goes through a black hole it enters a different dimension, namely the accompanying universe.

Note that, the matter particles reaching a black hole join the ones already there and cause the black hole to grow in size and become a wider gate for aether to flow through.

- **Light and other electromagnetic waves** are phase vibrations in aether. Light travels through aether just as sound does in a medium such as air or water.

    In this case, aether is not encouraged to have any kind of translational motion but it is simply made to vibrate locally, just like the waves formed on the surface of water in a pond as a pebble is dropped into it.

Aether versus Water or Air

Aether can best be represented as the water that is in an ocean, the ocean being the space. It can also be represented as the air that is in the atmosphere and the atmosphere being the space. Therefore,

- **Space,** is analogous to an ocean where aether has replaced the water. The essence of an ocean is its being filled with water. Therefore, if all of the water is drained out of a given ocean, then that ocean can no longer be called an ocean.

    The very same is true about space and aether. Space is filled with aether and is literally dependent on its presence. Therefore,

    **"The relation between aether and space is
    just like that of water and ocean."**

    In other words,

    "Space without aether is literally meaningless."

- **Time** is experienced by any object or any living being based on its motion relative to the local aether medium. This effect can be experienced in two ways:

    1- The object is moving in aether, or
    2- Aether is somehow encouraged to pass by the object.

In the first scenario, as the object speeds up, it experiences time at slower and slower paces. And, if it manages to reach the speed of phase vibrations in that medium, it will literally catch up with the phase vibrations that it has generated.

At this point, the object will no longer experience the passage of time. Even though it sounds strange, but one can say that,

"Time is actually progressing at the speed of phase vibrations in aether."

Therefore,

"Any object that moves at the same speed as the speed of phase vibrations in aether, will automatically be in sync with its own past/present, concurrently."

Imagine a supersonic airplane flying at the speed of sound in the air. As the airplane is travelling at exactly the speed of sound, using a very strong speaker, the pilot can broadcast his own voice counting 1, 2, 3, …., 100,….

Since the airplane is moving at the speed of sound in air, the portion of the sound waves, corresponding to each and every number stated, that are travelling towards the front of the airplane will be literally superimposed on each other, as if they are all stated at the very same time.

In other words, all of the numbers that were stated up to and including the one stated currently at the present will be heard simultaneously. Basically, the past starting from when the airplane reached and stabilized its speed at exactly the speed of sound in air, and present will be experienced at the very same time.

The drawing below demonstrates the sound waves generated during such a scenario.

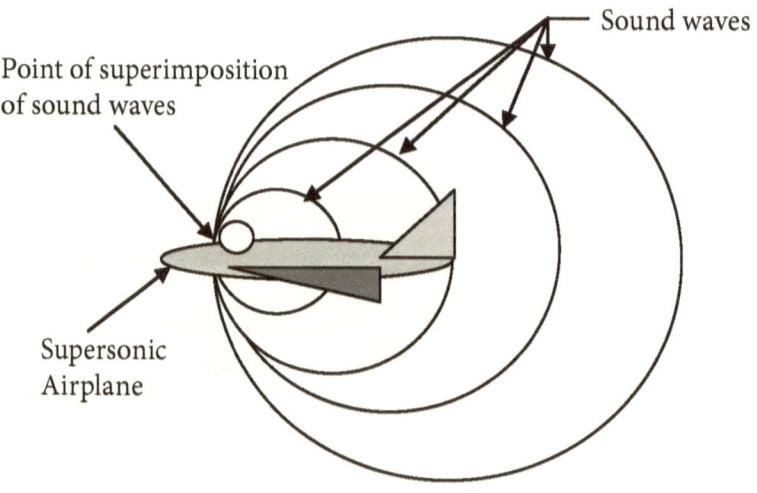

If the pilot decides to fly his airplane faster than the speed of sound, then the sound waves generated due to numbers stated in the past will literally fall behind and only the number that is currently stated will be heard.

The drawing below demonstrates the sound waves generated during such a scenario.

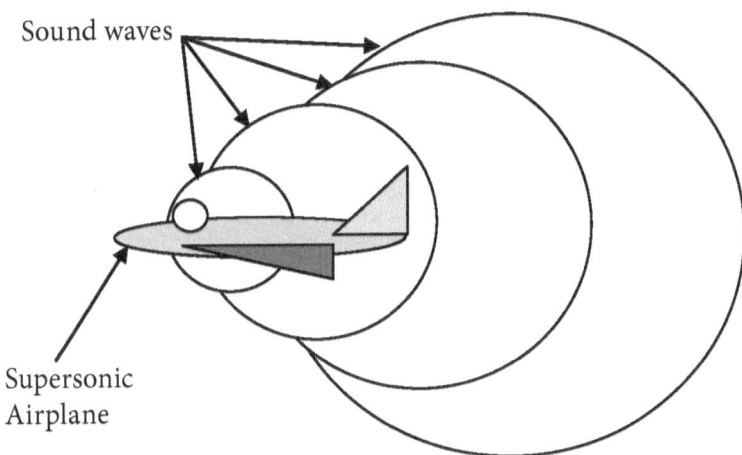

Sound waves

Supersonic
Airplane

Note that, just because the airplane is moving faster than the speed of sound in air, it does not imply that the pilot will hear the sounds corresponding to any of the numbers that are not stated, yet. He will hear those numbers, one by one, as they will be stated, in their own proper timing. In other words,

"The 'future' can only be experienced, as it arrives and becomes the 'present'."

Correspondingly, if the pilot decides to slow his airplane speed down to less than the speed of sound, as he is continuing with his counting exercise, the airplane will be falling behind from the wave fronts generated by the sounds. Therefore, he will be literally falling behind from the real present that he could be at.

The drawing below demonstrates the sound waves generated during such a scenario.

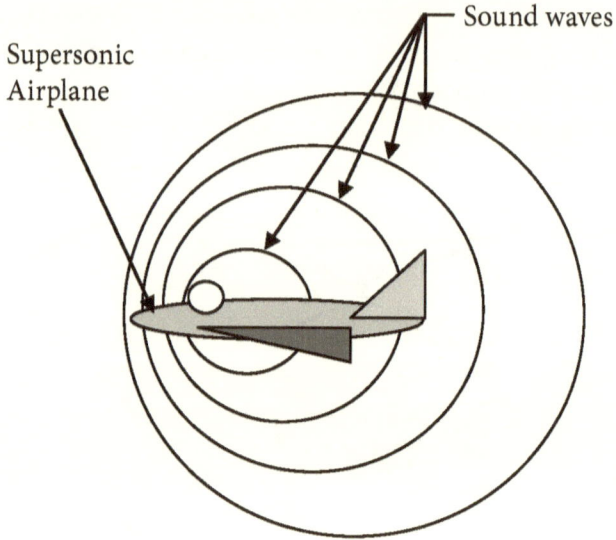

Supersonic Airplane

Sound waves

In other cases, where the object is kept relatively stationary and aether is made to pass by, either due to gravity, magnetic, or electric phenomena, it is as if the airplane is mounted stationary inside a wind tunnel and air is encouraged to pass by at various speeds.

The effects experienced due to increasing the speed of the wind generated by the wind tunnel up to and beyond the speed of sound will be exactly the same as those stated earlier, regarding the cases where the airplane itself was flying in the air.

In short, one could say that any object that is moving at a speed that is slower than the speed of the phase vibrations in aether, is literally falling behind from the wave fronts associated with its own past. Therefore, in normal circumstances,

> **"Objects and living beings experience their delay
> in time and not the passage of time itself."**

In other words,

> **"Objects and living beings are in fact experiencing time in
> reverse, as they are continuously literally falling behind from
> the real current time (present) that they could and should be at."**

It must be particularly noted that,

"The passage of time has nothing to do with space, as in being part of a 4-D space-time manifold of any sort. Time is only a side effect of phase vibrations in aether."

- **Energy** is like the variety of motions, such as translational or vibrational motions, that is manifested by the individual constituents (atoms/molecules/particles) forming the water medium.

- **Dark Matter** can be likened to the water itself (in its fluid state).

- **Matter** (as well as anti-matter) can be likened to the icicles that form in water. Even though icicles forming in the liquid water are less dense as compared to liquid water, but they are different states of the same entity, namely water.

- **Gravity** is like the drag force generated by the one way flow of water towards tiny drain holes that are all over the place and allow the water to go through, as if it is entering a parallel container (or an accompanying ocean).

- **Magnetic field** is like the mid-range activities such as a giant whirlpool which forms giant doughnut shaped flow patterns in the medium of water.

- **Electric field** is like the short range disturbances such as the local motions generated in water by a paddle.

- **Light,** being a phase vibration in aether, can be represented as the sound waves that are travelling in water or air. The speed of phase vibrations is constant, independent of the relative motions of the producer and receiver, and also acts as a barrier for objects that intend to go faster in that medium. But, it is breakable, given that the proper requirements are met, just like the speed of sound in air was viewed as a barrier and yet objects such as rockets or supersonic airplanes can go through it.

- **Black holes** can be viewed as being giant drain holes at the bottom of the ocean, where if deep enough, due to high enough a pressure and the proper size of the opening, the free flow of water going through will be at or higher than the speed of sound (phase vibrations) in water.

  In this case, even if someone, who is being dragged into such an opening, attempts to broadcast an, S.O.S. signal with the strongest possible speakers, his/her cry for help will not be able to travel away from the drain hole.

  Therefore, that region of the ocean can be referred to as the "**Silent Zone**", because no sound waves will ever be heard coming out of it.

Properties and Characteristics of Aether

Some of aether's properties are similar to those of regular matter and yet other properties are quite unique to aether itself. Properties and characteristics of aether are described below.

1- Aether cannot be created or destroyed.

The amount of aether that exists in this universe (and in the accompanying universe) is fixed.

Aether is a unique medium that basically acts as not only the carrier of a variety of forces, but also as the medium in which literally everything can exist, as it also gives meaning to space itself.

In the beginning, there was only aether. It was in a very compressed state. It experienced a sudden expansion in all directions. As it expanded and its density/pressure was reduced, under different circumstances it started to manifest in a variety of forms such as matter, anti-matter, dark matter and so on.

This process has been the same as the formation of icicles in water as its temperature is reduced to a certain level.

In other words, aether is not created and cannot be destroyed, however, under different circumstances, it does manifest as being in different states, just as water can manifest as ice, vapor or fluid.

Over time, aether that is in this universe is shared with another universe, namely the accompanying universe. But, the overall amount of aether remains unchanged.

2- Aether is a continuous and fluid medium.

Aether is a continuous medium that is literally freely flowing everywhere in this universe. It can be likened to a thick cloud that is encompassing everything. It does possess somewhat different local densities/pressures, but not by much since it is always freely flowing from the higher density/pressure regions to the lower ones.

3- Aether is a compressible/elastic fluid.

Aether is a compressible medium that can be likened to a gaseous medium. However, its compressibility is not due to its being made of individual particles that can be squeezed together, as pressure is raised, but rather due to its inherent elasticity.

4- Aether is a dynamic medium, even on the smallest scales.

Aether is not a stationary medium as it was assumed by the 19th century physicists such as Mr. Lorentz and Mr. Maxwell. It is quite a dynamic medium.

The force of gravity associated with individual celestial bodies both near and far, as well as the forces associated with the magnetic and the electric fields encourage the formation of a variety of flow patterns in the medium of aether, from the most microscopic scales to the most macroscopic scales.

These flow patterns in aether are just like the variety of flow patterns that exist in air in the atmosphere or water in the oceans.

5- Aether's viscosity is essentially nonexistent.

Aether's flow is always continuous and due to its next to nil viscosity, the density/pressure gradients between neighboring regions are quite gradual.

However, these gradual density/pressure gradients are detectable, on large scales, when the internal regions of galaxies or even individual solar systems are compared to their respective surrounding regions.

6- Aether is under extreme pressure.

Aether medium in this universe is under tremendous pressure. In fact, aether's flow towards matter particles and even black holes is pressure driven and not suction driven. The flow of aether through matter particles or black holes is just like the flow of air that is inside a high pressure air tank and is allowed to flow/escape through a valve as it is opened.

The local pressure of aether directly affects its local density since aether is compressible. Even though these density/pressure variations, between neighboring regions of space are minimal due to aether's lack of viscosity, they still encourage the flow of aether towards any available opening that may be near or far away.

Note that, the pressure difference between the aether that is in this universe and the aether that is in the accompanying universe can be calculated. See the section on "Predictions / Applications" for details.

7- Aether can flow at sub-light as well as super-light speeds.

Aether flows towards matter particles at sub-light speeds, since the size of such drain holes are too small to allow unrestricted flow of aether through. However, aether does speed up to the speed of light as it reaches the event horizon of any black hole, and as it continues past the event horizon, its speed enters the super-light speed regime.

Note that, aether's speed at the event horizon of any black hole is exactly equivalent to the speed of phase vibrations (speed of light or other electromagnetic waves) in the local aether that exists at that particular black hole's event horizon.

8- Aether has energy but no mass.

Aether is not made of any kind of particles. It does not have any mass associated with it. However, it does possess energy. In fact, it possesses a lot of energy which is due to:

1- Its constant motion, kinetic energy, and
2- Its internal pressure, potential energy.

Mass is literally nonexistent in this universe. Mass is only a manifestation of aether in its condensed form, as matter and anti-matter particles are formed due to spikes/resonances in the phase vibrations in aether. More details are provided below.

9- Aether is not directly affected by temperature.

Temperature is a phenomenon that affects matter and anti-matter. It does not directly affect aether in its normal fluid state. It does, however, affect aether indirectly since higher temperatures vary the way the matter content of this universe behaves.

For instance, higher temperatures lead to general expansion of the local matter density which directly implies less number of drain holes per unit volume of space for aether to flow through.

10- Aether formed the initial container (space) for this universe.

Aether was the first thing that was formed in this universe, at the beginning of time. Even 'time' started its existence as aether came into existence and started hosting phase vibrations.

Aether basically became the medium or the host for everything else such as matter, anti-matter and even electromagnetic waves which are all literally different manifestations of aether itself. In other words,

> **"Aether can exist independently of any and all other apparent contents of this universe, while the very existence of aether is a prerequisite for the existence of everything else."**

11- The relation between aether and matter and anti-matter particles.

Phase vibrations in the medium of aether are just like regular sine waves. They do have a positive half and a negative half, during every complete cycle.

Under certain conditions, phase vibrations in the aether medium can be encouraged to form full wave or even half wave spikes and so lead to the formation of condensed states in that medium.

When spiked (condensed), the positive halves lead to the formation of particles that are of positive (+) charge and the negative halves lead to the formation of particles that are of negative (-) charge.

The matter and anti-matter particles that exist in this universe are literally condensed forms of aether that are floating in the aether medium.

Simple presentations for both full wave and half wave spikes are shown below.

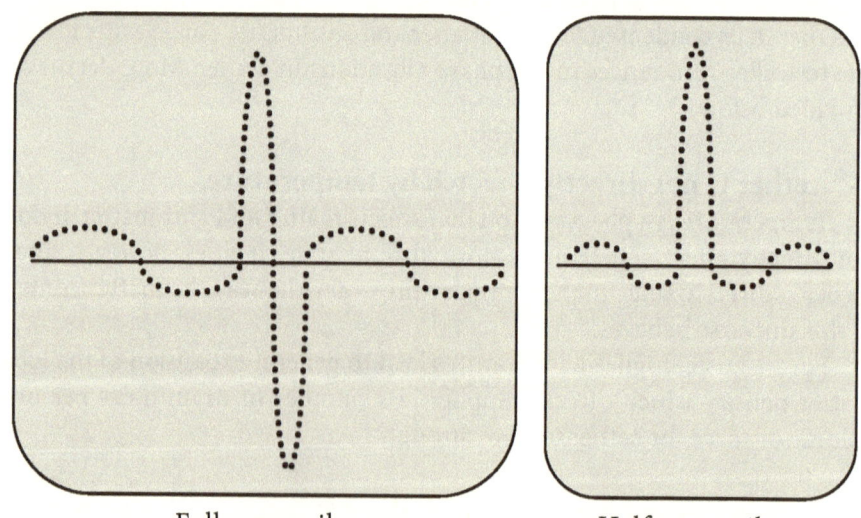

Full wave spike   Half wave spikes

This is almost like the formation of ice in water, with the exception that ice is less dense as compared to water, while condensed states of aether corresponding to matter and anti-matter particles are much denser than aether in its normal fluid state.

According to the currently accepted definitions in particle physics, the only difference between matter particles and their anti-matter counterparts is their electric charges. For example, a proton is positively charged (+), while an anti-proton which weighs the same as proton is negatively charged(-).

However, as far as the relations between aether and matter and anti-matter particles are considered, all particles of positive charge are grouped together and all particles of negative charge are grouped together, regardless of their belonging to the normally defined matter or anti-matter particle families.

Therefore, according to this newly proposed matter particle categorization strategy, protons (+) and anti-electrons or positrons (+) are grouped together, just as anti-protons (-) and electrons (-) are grouped together.

Therefore, as the full wave (a complete sine wave cycle) is spiked (condensed), the result is the formation of a pair of particles that are of opposite electric charges. Correspondingly, as a half wave (a half sine wave cycle) is spiked (condensed), the result is the formation of a single particle of either positive or negative in charge, depending on the half wave being positive or negative.

Note that, the types of particles formed during these processes solely depend on the size of the spike formed. For example, the more pronounced spikes give birth to proton and anti-proton pairs, while the weaker spikes can only promote the formation of electron and positron pairs, and so on. In other words,

"Density spikes in the medium of aether manifest themselves as matter and/or anti-matter particles."

Note that, the neutral particles, such as neutrons, where formed from the joining of electrons and protons (also Anti-protons and positrons), as they attracted each other due to having unlike charges.

Since the phase vibrations were distributed uniformly in the aether medium, the same circumstance was provided throughout its volume. Therefore, the phase vibrations present in aether resulted in the formation of matter and anti-matter particles everywhere, at the same time. In other words, one could say that,

"The whole universe literally bloomed across its volume with matter and anti-matter particles, at about the same time."

12- Aether and the cosmic background radiation

As most of the matter and the anti-matter particles reunited and annihilated each other, their wave forms were literally spread/ dispersed in aether as lower amplitude phase vibrations.

These phase vibrations were further spread and flattened by the very expansion of the aether medium. Over time, these phase vibrations spread everywhere and literally became a uniform background noise in aether.

At the present time, these weak phase vibrations are detected as **cosmic background radiation** that exists throughout the whole universe.

13- Matter-Energy equivalence,

Any matter particle is in fact a spike (a condensed state) in aether. If for any reason a given matter particle pair or a single matter particle (a full wave or a half wave spike, respectively) is to be transformed into pure energy which is actually of the electromagnetic type (phase vibrations), it basically implies that the spike corresponding to that particle(s) is to be flattened out and its volume is to be spread into its surrounding medium.

In other words, the condensed aether that is the particle(s) is to be dispersed in the normal aether in fluid state that is forming its surroundings. This dispersion process of matter/anti-matter spikes into energy can be likened to the sudden melting of an iceberg as it is floating in the ocean. This very action causes the formation of a temporary high local fluid aether density that tends to spread into its surroundings. In doing so, it generates an intense phase vibration. The following figures show such a transition in a very simple way.

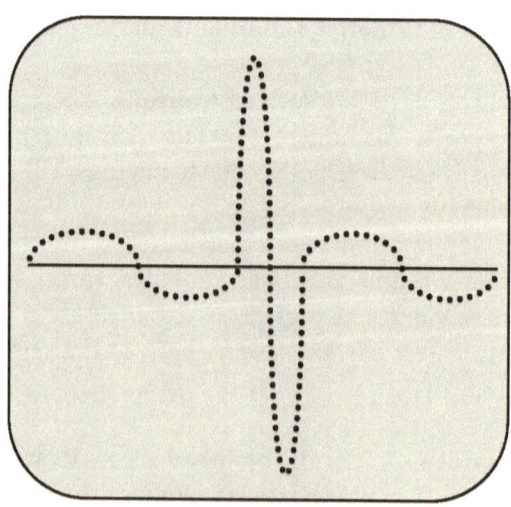

Before Annihilation

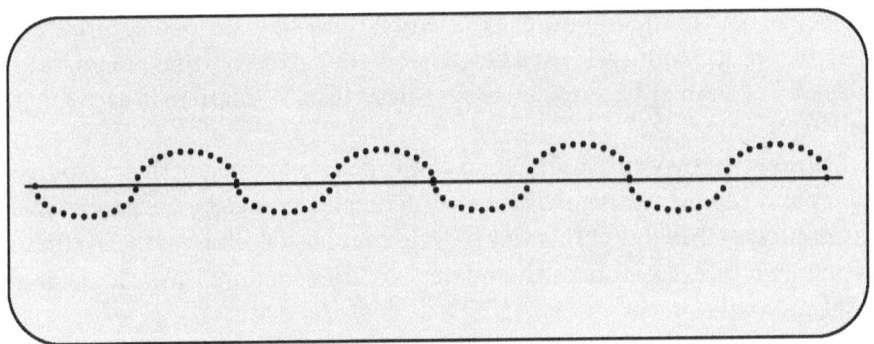

After Annihilation

In other words,

"The initial matter and the resulting energy are different manifestations of literally the very same amount of aether that was once condensed into a spike form and then is flattened and eventually dispersed in the general medium of aether."

If many particles go through such a process simultaneously, they will cause an expanding shock wave in the local aether medium. This shock wave would literally carry with it anything and everything in its path. This process is almost like what takes place in a nuclear explosion.

Note that, in the case of nuclear explosions, only the binding energies between some of the nucleons are released. The overall number of nucleons (protons and neutrons) is still conserved, even though quite a few of the nucleons will change their identities from proton to neutron or from neutron to proton, during this process.

14- Aether and the formation of stable vs. unstable matter particles

If the matter particles (spikes or resonances formed in the phase vibrations in aether) are isolated and do not encounter other spikes that are of opposite type (out of phase) they will be stable particles. However, if they encounter other spikes, they will face challenges in staying as they are, and eventually will lose their identities. These particles manifest as being unstable particles.

15- Relation between aether and radioactive isotopes

Every matter particle is in fact a manifestation of a spike (condensed state) in the medium of aether. As each of these matter particles chases its respective path in the volume of a nucleus, it causes a disturbance/turbulence in aether, in its vicinity. If particles generate disturbances that are too far or mismatched in their <u>frequencies</u>, they will not usually interfere in each others' well being.

However, if they are close enough and/or the frequencies of their respective spikes and hence the disturbances they generate are matched, they can form harmonics, as they fly by. This way they do cause instabilities in the livelihood of one, two or even more of the matter particles that are involved. In most cases they will:

- Lose part of their energy as a newly formed phase vibration that is broadcast in the surrounding medium, (electromagnetic wave, as in gamma rays) or

- Divide into smaller spikes (beta decay) or

- Get knocked off of their path within the nucleus, either individually or as a collective (alpha decay or total splitting of the nucleus).

Either way, the overall process leads to a change in the identity of the nucleus, as a whole, and a new isotope or even new elements are formed.

16- Relation between aether and the half-life of radioactive isotopes

The matter particles in the nucleus are in orbit around each other and they are literally continually repeating a sequence of flyby routines that are apparently in random order. Every so often during their flyby passes they encounter a rare occasion when their generated disturbances become in sync and create harmonics that are strong enough to cause one or more matter particles to:

- Lose part of their energy as a newly formed phase vibration that is broadcast in the surrounding medium, (electromagnetic wave, as in gamma rays) or

- Divide into smaller spikes (beta decay) or

- Get knocked off of their path within the nucleus, either individually or as a collective (alpha decay or total splitting of the nucleus).

Such occurrences are not based on chance, since the motions associated with each and every matter particle is governed by certain forces that they encounter along the way. Some of these forces are caused by the other members of the same nucleus, while some others originate from the outside. The external forces are due to the neighboring nuclei as well as the other particles that are freely wandering in the space between the nuclei.

In other words, the occurrence of nuclei disintegrations can be viewed as if they are totally based on chance, yet they are not. Their behavior is exactly like flipping a coin. If it is done only once, the outcome is uncertain until the coin is actually thrown and comes to rest. This scenario corresponds to considering a specific nucleus. Its disintegration cannot be readily predicted until it actually takes place.

However, if the very same coin is thrown a multitude of times, the outcomes will average around 50/50, since there are only two possible outcomes. Such an outcome is simply due to the existence of too many variables that affect the motion of the coin in its journey during each throw until it comes to rest. Therefore, the outcome evenly averages between the two possible outcomes and the overall process seems to be based on chance, yet it is not.

Since, if one can repeat the very same effects, during every throw, over and over, the results will always be identical. In this case, the results can be all heads or all tails, as desired.

Similarly, the nuclei that are present in any radioactive sample experience occasions when their own constituents (as well as external particles and other nuclei) generate disturbances that are strong enough to cause the formation of harmonics. There are literally millions of millions of millions of nuclei present even in a single gram of any kind of radioactive isotope. Therefore, the occurrences of such harmonics among all nuclei that are present, averages out.

However, when one specific nucleus is considered, if the state of every particle within the nucleus is known, one can calculate the exact timing of when that nucleus will encounter that rare occasion when the disturbances caused by its constituents will generate strong enough harmonics that will change the identity of that nucleus.

Note that, shorter and longer half-lives are simply due to how often the needed circumstance occurs in a given nucleus that can lead to the formation of a strong enough harmonic. If not influenced externally, the nucleus will experience the needed circumstance after a specific number of flybys of its constituents and will disintegrate, regardless of how many other nuclei may be present in the same sample.

17- The relation between aether and dark matter, dark energy and vacuum energy

The relation between aether and dark matter, dark energy and vacuum energy can be specifically described as follows:

- **Aether's relation with dark matter**
  The aether that exists everywhere in this universe, in a compressible fluid state, is the true form of dark matter.

  Note that, in certain conditions, as explained earlier, as phase vibrations in aether form spikes, dark matter becomes what is known as the regular matter.

- **Aether's relation with dark energy**
  The energy associated with any and all forms of phase vibrations that are carried in aether is the true form of dark energy.

  Note that, the whole range of the known spectrum of electromagnetic waves as a class of waves is only one of many types of phase vibrations that exist in aether.

- **Aether's relation with vacuum energy**
  The potential energy associated with aether's internal pressure which is the driving force for its expansion, is the true form of vacuum energy.

  Note that, the very existence of black holes in this universe implies that aether's internal pressure in this universe is still very high, as compared to the aether's pressure in the accompanying universe. Therefore, one can say that,

The vacuum energy that is associated with the internal pressure of aether overshadows the other forms of energy present in this universe.

As the regular matter and anti-matter are slowing down the expansion process of this universe, aether's internal pressure is trying to speed up this very expansion process. Therefore, one can say that,

**Aether concurrently has contradictory influences on
the expansion process of this universe, as a whole.**

This contradictory effect can be better understood by visualizing two giant magnets that are freely floating in deep space, with their North poles pointing at each other. In this case, while their gravitational forces will pull them towards each other, their repulsive magnetic forces will push them apart.

18- How does an object's relativistic speed, relative to its local aether medium, affect its physical dimensions?

According to Mr. Lorentz's transformation equations, as an object's speed approaches the speed of light in a vacuum, its dimension in the direction of its motion contracts. Theoretically, if an object can manage to reach the speed of light, it will become totally flat in its geometry.

In order to understand why such an effect takes place and why the object is only affected in the direction of motion, one needs to examine the object on a microscopic level, namely the atomic level. The atoms in some objects are arranged in an organized fashion and form a lattice structure of some sort. In other object's atoms do not follow any particular pattern in their spacing. Each individual atom is also a 3-D entity which includes a central nucleus that is surrounded by an electron cloud.

As the speed of an object approaches the speed of light in a vacuum, the atomic vibrations and the overall geometries of the atoms and the nuclei gradually change, but only in the direction of motion relative to aether. They change due to the drag force that they receive from the aether that is rushing through their composition.

Note that, the dimensions of the object along the other two directions (that are perpendicular to the direction of motion) are not affected by the speed of the object, as it approaches the speed of light in a vacuum. This is due to the fact that there is no change in the drag/push that is experienced from the sides.

In other words, as the object speeds up, its components mandatorily respond to the changes that occur in the aether's flow in the direction of motion. The following figures demonstrate this compliance by the individual constituents on different levels, as the speed of an object with a crystalline lattice structure approaches the speed of light (the speed of phase vibrations in the local aether medium). The same can also be visualized for an object that consists of atoms that are not well organized.

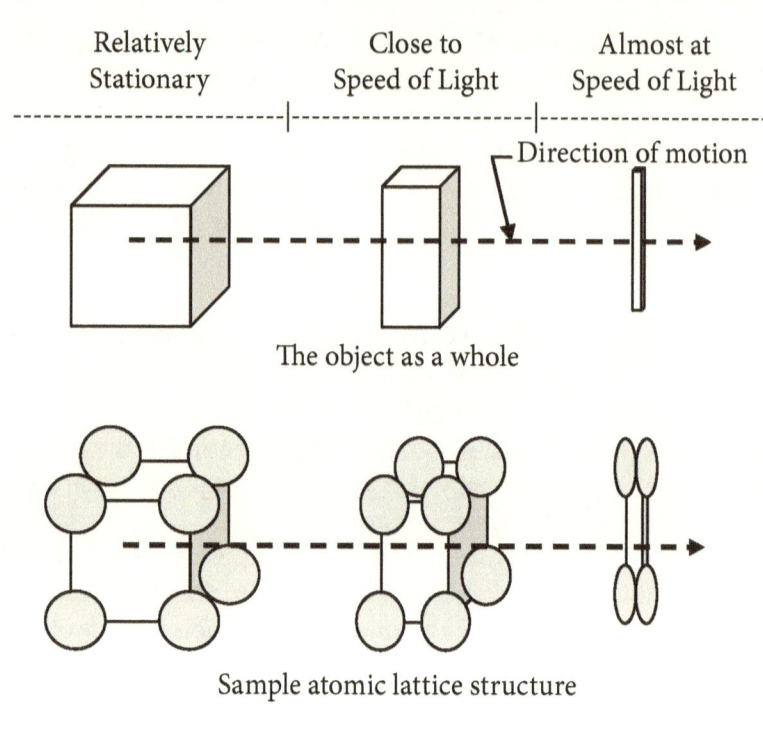

The object as a whole

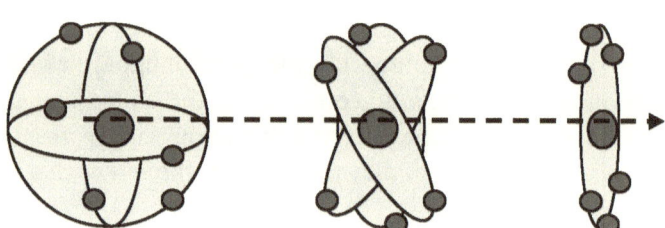

Sample atomic lattice structure

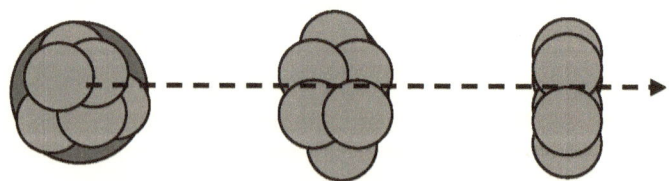

The individual atoms

The individual nuclei and their nucleons

Note that, the individual subatomic particles do not get squeezed in the direction of motion.

Note that, if an object stands relatively stationary and the medium of aether is encouraged to pass by it, the effect will be just the same. For example, if an object is placed in a strong gravitational field, a strong magnetic field or a strong electric field, it will experience the very same effects on its longitudinal dimension (that is in the direction of the flow that is induced in aether) as if it were travelling at high rates of speeds in aether. In other words,

"The relativistic effects on the physical dimensions of an object depend on that object's motion relative to its local aether medium."

19- The relation between mass and density of an object and its speed relative to its local aether medium

As an object's speed approaches the speed of light in a vacuum, it contracts in the direction of its motion. However, the other two dimensions, that are perpendicular to its direction of motion, remain unchanged.

Therefore, as the speed of an object approaches that of light (the speed of phase vibrations in the local aether medium), its volume decreases linearly with its length, that is in the direction of motion. This automatically implies that, not the mass, but the density of the object increases as its length shrinks. In other words,

"As an object speeds up and approaches the speed of light in a vacuum, its density approaches infinity and not its mass. Its mass remains unchanged."

The energy that is utilized to increase the object's speed relative to its local aether medium is in fact stored in the object as different forms of potential energies. These stored potential energies are shared between various components of the object, as they are literally compressed towards each other and flattened out of their normal shapes/positions. Just like the energy that is stored in a spring as it is compressed, or as the air inside a cylinder is squeezed in one particular direction, by pushing on the piston.

Note that, as an object's speed approaches that of the phase vibrations in the local aether, even though its mass stays unchanged, but its weight increases, just as if it is experiencing a stronger gravitational field. This is due to the fact that, in both cases, the aether is rushing by at higher speeds. In other words,

If an object manages to reach the speed of light in a vacuum, it will literally experience the very same effect that it would if it were at the event horizon of a black hole, where the speed of aether rushing by is exactly equivalent to that of the phase vibrations in that medium.

The figures provided in the previous section clearly demonstrate the variety of deformations that are induced in the internal structure of an object, as its speed relative to its local aether approaches the speed of the phase vibrations in that medium. These deformations are literally due to the push/drag that is exerted by aether as it is made to rush by, either because the object is moving in aether or aether is encouraged to move relative to the object.

Note that, if an object stands relatively stationary and the aether is encouraged to pass by it, the effect will be just the same. For example, if an object is placed in a strong gravitational field, a strong magnetic field or a strong electric field, it will experience the very same effects on its density as if it were travelling at high rates of speeds in aether. In other words,

"The relativistic effects on density and weight of an object depend on that object's motion relative to its local aether medium."

20-Aether's overall density/pressure is decreasing

The overall density/pressure of the aether medium is gradually decreasing with time, due to two independent processes:

- **The expansion of the universe which is the aether medium.**
  As the universe is expanding, the volume of space which is the very same as the volume of aether, is also increasing. Therefore, as the universe expands automatically the overall density/pressure of aether is reduced.

- **The leakage of aether out of this universe.**
  As aether is continually leaking from this universe, by going through matter and anti-matter particles as well as black holes, the overall aether content of this universe is gradually reduced. This process automatically lowers the overall density/pressure of aether in this universe.

Note that, the aether that escapes from this universe through matter and anti-matter particles or black holes is not lost; it literally leaks into the accompanying universe.

21- Changes in the strength of the force of gravity as a function of distance

The strength of the force of gravity is dependent on the distance to the gravitating object (planet, star or even a single particle). This relationship was given by Newton and is known as the inverse squared law.

More accurate measurements will indicate that the force of gravity associated with any aggregate of matter particles (within a galaxy), changes faster than the quadruple, at half the distance. This is due to reductions in the aether density as it approaches a large aggregate of matter particles, such as a star. In other words,

On smaller scales (within galaxies), the force of gravity changes a little bit faster than the inverse squared law. or,

$$F \approx Gm_1m_2 \, / \, x^{2.001}$$

Also, the separation of galaxies is changing due to the expansion of the aether medium (space) as a whole. This effect leads to an apparently weaker force of gravity than expected over long distances. This is due to the fact that more distant galaxies are observed as they were a long time ago, when aether was also much denser. In other words,

On larger scales (intergalactic), the force of gravity will seem to change slower than the inverse squared law predicts. or,

$$F \approx Gm_1m_2 \, / \, x^{1.999}$$

22- The force of gravity is gradually weakening.

The force of gravity which is in fact the drag force induced by the flow of aether is directly dependent on the aether's density. Therefore, as the density of the aether medium is reduced, due to expansion and leakage, its flow is becoming less effective in inducing the force of gravity.

23- Widening of the planetary orbital paths due to the decrease in the local aether density

As the aether's density is decreasing over time, due to its expansion and leakage, it is becoming less effective in dragging objects that happen to be in its path. Therefore, stars are gradually exerting less of a gravitational force on their respective planets. This simply means that stars are gradually losing their grip on their planets and consequently

"The planetary orbits are gradually becoming wider."

This side effect of decrease in aether's density has already been detected, in regards to the planet earth. According to the collected data, the orbital path of the earth is widening by about 7 meters (about 23 feet) per century.

Of course, a small portion is justifiably due to the sun losing mass as it is continuously transforming some matter into energy and also as it is literally throwing some matter into space as solar storms.

24- Expansion rate of the universe seems to be speeding up rather than slowing down.

The expansion process of the universe, which is based on direct observations, is graphically presented below.

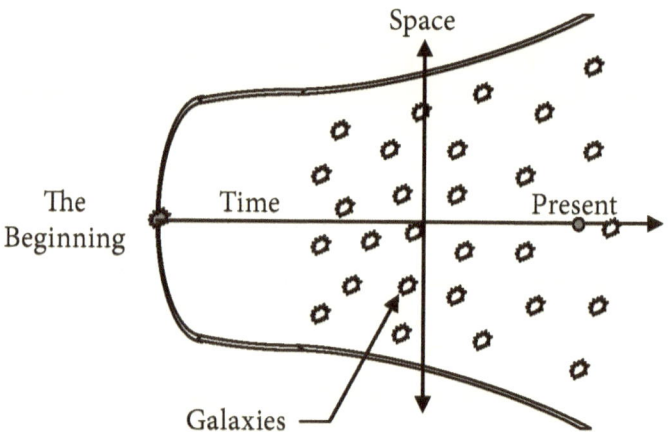

According to these findings, the universe seems to be accelerating in its expansion process, rather than decelerating as it is expected based on the known laws of gravity.

This is due to the fact that the overall force of gravity is gradually becoming less effective due to two different reasons:

1- The expansion of the universe which is literally increasing the separation of the galaxies from each other. The farther the galaxies get from each other, the weaker they affect each other gravitationally.

2- The reduction in aether density due to both its overall expansion as well as its leakage out of this universe. Less dense aether is correspondingly less effective in inducing a drag force, as it flows.

These two reasons directly influence the effectiveness of the force of gravity in slowing down the overall expansion rate of this universe.

Also, when considering long distances, such as billions of light years, due to longer time spans involved, some other factors also come into play. These factors are:

1- The gradual increase in the speed of light (speed of the phase vibrations in aether), which is due to aether density gradually decreasing,

2- The gradual increase in the pace at which time is experienced, which is due to increase in the speed of the phase vibrations in aether while the relative motions of celestial bodies are not changing as much.

When these effects are also taken into account, the observed behavior can be readily explained.

Suppose two galaxies, one located at about one billion light years and the other about one half of one billion light years away from earth. They can be detected to be receding from earth at say about 1,000,000 km/hr and 600,000 km/hr, respectively. The observed discrepancy is not due to the universe accelerating in its expansion, but rather due to the speed detected for the farther away galaxy. That galaxy's speed should be corrected for the variations that have occurred during the past one billion years.

Once the proper corrections have been implemented, it will become clear that the speed measured for the farther away galaxy is actually much higher than it seems to be at the present. In other words,

**The variations in the 'speed of light' and the duration of 'time'
are creating such an illusion that the universe is speeding
up in its expansion, while it is actually slowing than.**

25-Esmailzadeh Critical Aether Pressure Difference

At a certain point in the expansion process of this universe, the difference between the aether pressure in this universe and the aether pressure in the accompanying universe will approach a critical value. That value of the pressure difference corresponds to the minimum required pressure difference that would cause the flow of aether through the largest possible black holes to reach the speed of light (speed of the phase vibrations in aether). This critical aether pressure difference can be referred to as:

"Esmailzadeh Critical Aether Pressure Difference"

In other words, once this critical value of the aether pressure difference between this universe and its accompanying universe has been reached, all of the black holes, regardless of their sizes, will lose their status as black holes. This is due to the fact that, the speed of the aether reaching their surfaces will be just under the speed of light. This in turn means that even though black holes will keep on devouring more planets and stars, but light and other electromagnetic waves will be able to travel against the incoming flow of aether and escape.

Note that, two factors affect how fast/soon this critical aether pressure difference will be reached:

1- The reduction in the overall pressure of aether in this universe, since it is the driving force to push aether through any possible opening.

2- The increase in the speed of the phase vibrations in aether as the aether is becoming less dense with time, since this speed is also the speed at which light and other electromagnetic waves will be travelling.

Therefore, as time goes on, the aether pressure in this universe is reduced while the speed of light and other phase vibrations in aether are increased. At some point in time, it would just not be possible for the existing aether pressure difference between the two universes to cause the flow of aether through the biggest openings (black holes) to reach the speed of light. Encountering such a circumstance will serve as the clear indication that the *Esmailzadeh Critical Aether Pressure Difference'* has been reached.

26-All of the black holes will gradually lose their status as black holes.

As the Esmailzadeh critical aether pressure difference is approached, all of the black holes, starting with the smallest ones, will lose their status as being black holes. At this point, the local aether pressure difference will no longer be sufficient to cause the aether flow to reach the speed of light in that medium, before it reaches the surface of the black holes.

This simply implies that the black hole will no longer possess an event horizon, because light and other electromagnetic waves will be able to escape their grips. In other words,

"Eventually, all of the black holes will become visible."

27- The dismantling of the galaxies, solar systems, stars and even planets and so on

As the force of gravity gets weaker and weaker, gradually, the following will take place all over this universe:

1- Stars will gradually let their respective planets widen their orbital paths. Galaxies will also grow in size, due to their stars following wider orbits.

2- Even before the force of gravity reaches its total ineffectiveness, stars and planets that used to form solar systems, will no longer have any desire (the needed gravitational attraction) to stay together.

    At this stage, not only the solar systems but also all of the galaxies will lose their structural integrities. As a result, all of the celestial bodies will act just as independent atoms do in a gaseous medium. In so doing, they will literally float everywhere in this universe, without having any preference or desire to go towards any particular direction. They would simply follow their momentums.

3- During later stages, even the stars and planets will lose their structural integrities, since their gravitational forces will not be strong enough to hold them together any longer.

At this point, the atoms and molecules forming stars and planets will literally let go of their unions with each other and freely float in the vast expanse of space.

28- The force of gravity will be neutralized, altogether.

Eventually, the pressure of aether in this universe will equalize the pressure of aether in the accompanying universe. Once this pressure is reached, the flow of aether towards matter particles will slow down to a complete stop. In other words, as the pressure difference between this universe and the accompanying universe approaches zero,

"The force of gravity will gradually fade away."

From this instant onwards,

The force of gravity will become none existent.

At that point in time, if the universe has any momentum left in its expansion, it will keep on expanding forever, since the universe would literally lack its braking instincts. This era in the existence of the universe can be referred to as,

"Esmailzadeh Zero Gravity Era"

29- The universe will never stop its expansion process

The volume of aether will always keep on expanding, even after its pressures in this universe and the accompanying universe have been equalized. This expansion process will literally continue until the universal density/pressure of aether reaches zero. This density/pressure will be realized only when the size of the universe reaches infinity.

30- The speed of light and other phase vibrations in aether are gradually increasing.

The speed of light and other types of phase vibrations in aether are dependent on the aether's density. Higher aether density means that the phase vibrations can travel slower in that medium.

Therefore, in the very beginning, when the density of aether was literally next to infinite, the speed of the phase vibrations (the speed of light) in that medium must have been next to nil. In other words,

In the beginning, during the sudden expansion of aether, the phase vibrations spread more effectively as they were carried by the expansion of aether rather than by travelling in it.

During the expansion process, as aether itself was expanding, it literally stretched the wavelengths of the phase vibrations that were present in it.

As the density/pressure of aether was reduced, due to its expansion, the speed of the phase vibrations in that medium was increased. The very same process is still taking place. In other words,

"As the density/pressure of aether is reduced, the speed of light (the speed of the phase vibrations in that medium) is increased."

Since aether's density is approaching zero, the speed of the phase vibrations in that medium will correspondingly approach infinity. In other words,

"The speed of light and other phase vibrations will ultimately reach infinity."

Note that, even though light and other electromagnetic waves are limited to speeds that are allowed for phase vibrations in the local aether medium, objects or living beings are not limited by such restrictions.

31- Time is gradually experienced at an increasingly faster pace.

The rate at which the passage of time is experienced by any object or living being is directly dependent on the speed of that object relative to aether, as compared to the speed of the phase vibrations in that medium. Therefore, as the speed of the phase vibrations in aether is increased due to reduction in aether's density, the time experienced by any and all objects, universally, is speeding up. In other words,

"The universal time is gradually speeding up."

Eventually, as the speed of phase vibrations in aether approaches infinity, the rate at which time is experienced by all objects in this universe will approach infinity, as well. In other words,

"Gradually, as time speeds up, the duration of minutes, hours, days, years, even millions of years will become equivalent to what is one 'second', at the present."

32- Aether can carry many different phase vibrations, as well as motions, concurrently.

In this regard, aether behaves just like air. Air can concurrently host countless number of phase vibrations such as sound waves that are at different frequencies. For example, this is commonly experienced as one hears a variety of sounds while walking down the street, particularly during busy hours, or as one listens to the live performance of a symphony orchestra.

Aether also allows concurrent transmission of many phase vibrations (light and other electromagnetic waves) with different frequencies through its composition. The very reception of various colors from different objects such as flowers, trees, buildings and so on, let alone the presence of millions of radio, TV and cell phone signals, which are all transmitted through the very same volume of space, as phase vibrations, clearly indicate aether's capability in performing such a task.

33- Aether does not directly interact with regular matter or anti-matter.

Aether, in its normal fluid state, is a neutral environment for anything and everything that exists in this universe. It is literally a host for all of this universe's contents and a carrier for the variety of forces through which they interact with each other. This is like air, since various objects such as airplanes use air to not only support themselves against the force of gravity but also propel themselves.

One can also say that, by its presence, aether literally enables the contents of this universe to move around and interact with each other (both near and far) in a variety of ways which they could not do otherwise. In short,

"Aether is not only the host for the contents of this universe but also the carrier of the various forces that are exchanged between them."

34- Aether can enter and exit this universe

Aether can enter and exit this universe, but only as it travels back and forth between this universe and its accompanying universe.

Note that, the accompanying universe is not a parallel universe, as in identical to this universe. The other universe is only an accompanying universe with 3 spatial dimensions. It has its own set of duties to perform.

35- Aether in the accompanying universe

In the accompanying universe, aether is at a much lower density/pressure as compared to this universe. Therefore, the following can be well expected as one may enter its environment:

- The speed of light and other electromagnetic type of waves is much higher in the accompanying universe as compare to what they are in this universe.

- The rate at which 'time' is experienced in the accompanying universe is at a much faster pace as compared to the rate at which it is experienced in this universe.

36-Aether cannot be detected directly.

The existence of aether can only be detected indirectly and through various tasks that it performs in regards to different affairs of the physical universe. For example:

- The very existence of matter and anti-matter particles is an indication of aether's presence in a condensed (spiked/ resonance) form.

- If the force of gravity is felt, it can only be due to the linear flow of the local aether towards some matter or anti-matter particle (s) that may be located anywhere in this universe.

- Any magnetic field is due to a flow generated in the local aether that is forming a small to large size loop.

- Any electric field is due to a flow generated in the local aether that is forming a small size loop.

- If any electromagnetic type of wave such as light is transferred through anywhere, it is automatically an indication that aether exists in that volume of space.

37- Aether is the only content of this universe.

Aether was the first content of this universe. Over time, and due to different circumstances, aether manifests its many faces which include various forms of matter and energy.

For instance, based on the info presented in this book, it is proposed that matter (including anti-matter) is only a condensed form of aether. Also, the electromagnetic waves in general are phase vibrations in the medium of aether. In other words,

"Everything in this universe including matter, anti-matter and dark matter, as well as all forms of energy, including dark energy and vacuum energy, are simply different manifestations of aether, the one and only content of this universe."

What is Space?

What is Space?

Space is measured and analyzed using units that are dependent on each other. For example, one kilometer is one thousand meters, or one light year is equivalent to the distance that light travels in one year, in a vacuum and so on. The only direct definition that one can propose for space is that,

"Space is the gap between objects."

To understand what space really is, one needs to take its various properties, as well as its relationships with other physical phenomena such as aether into account.

As they are explained in separate sections, in this book, many different phenomena in this universe are interdependent. To truly understand any of them, one has to have some understanding about the real essence of the others, as well.

This section is devoted to defining what "space" is, what its properties are, as well as how it relates to the other physical phenomena. The information in this section is presented in a way that allows one to concentrate on specific points.

1- Space is like an ocean that is filled with aether rather than water.

The relationship between space and aether is just like that of ocean and water. The existence of water literally gives meaning to an ocean. If water in an ocean is somehow made to expand, it will directly lead to an expansion of the ocean. And, if all of the water is taken out, the word ocean loses its meaning. The very same is true about aether and space. For space to exist, it must be occupied by aether.

Space is basically the container of everything in this universe. It primarily contains aether. The rest of the contents of this universe are literally floating in aether. The very existence of aether has given meaning to space, and the expansion of aether has lead to the current size of the universe.

Note that, space in fact encompasses this universe and its accompanying universe. The two universes are connected together through countless number of matter as well as anti-matter particles which allow the flow of aether from this universe to the other universe.

The aether that is in this universe is at a much higher pressure as compared to the aether that is in the accompanying universe. That is why, the flow of aether from this universe to the other one is pressure driven. This means that, matter particles actually act as pressure relief valves for the aether that is under high pressure in this universe, as they allow it to escape to the other universe.

2- Space is and will be expanding literally forever.

In the beginning, space was contained in a very small volume, because aether was in its super compressed state. Aether was in fact in a critically compressed state, to be more precise. Due to some phase vibration that was introduced into it, aether started a sudden expansion. As aether expanded, so did the vastness of the space which was occupied by aether. Over time, the space became the size it is observed to be, today. More details in this regard are provided in another section of this book, "Evolution of the Universe".

Space (and not just the physical universe) is still expanding, and will keep on expanding forever. This is due to the fact that, aether medium itself is expanding. However, aether's expansion rate, which is directly related to its pressure, is decreasing as time goes on.

Note that, the outer boundaries of space, or the front line of the aether's expansion is even farther ahead of the phase vibrations (electromagnetic waves such as light) that are spreading in that medium.

The expansion process of the matter contents of this universe is gradually slowing down, and it may come to a halt at some point in the future, so long as it does so before the force of gravity is neutralized. However, the expanding aether medium will keep on spreading the frontiers of space literally forever.

The expanding rate of aether will gradually reduce as its pressure in this universe is reduced. But, its expansion process will never come to a complete stop, since the pressure of aether can only approach zero. In other words,

"The radius of the space that is hosting this universe is and will be expanding at ever slower rates and yet it will never come to a complete halt."

3- Space is finite and it is spherical in its geometry.

The overall geometrical shape of space is spherical, since aether's expansion has been and is symmetric in all spatial directions. Aether's outer limits or its overall front line is a sphere which is expanding just like a balloon. This automatically indicates that the overall size of space is finite. One may ask:

<u>What is aether expanding in? Isn't that the true space?</u>

These questions are more philosophical questions than they are related to the physical universe. They are just like the questions that one can ask regarding the source of aether and what was the source of that source, and so on.

Any answers to such questions can only be based on pure speculation. This is due to the simple fact that, in order to even know what the outside of this universe (or even a house, for that matter) looks like, let alone where it is located, one needs to literally see it from the outside. Because, even if the whole of the interior of a house (for example) can be surveyed and precisely mapped out, still due to the lack of knowledge about the thickness of different exterior walls, one would not be able to even guess what the outside of the house may look like.

4- The internal structure of space is like a sponge.

The internal structure of space can be analyzed on two different levels:

- **On the microscopic scale,**
  On the microscopic scale, the internal structure of space is quite like a sponge which has countless number of tiny holes in it. These holes are in fact the matter particles that have formed due to the formation of spikes in the phase vibrations in aether. These holes literally act as drain holes through which aether is escaping from this universe into its accompanying universe.

- **On the macroscopic scale,**
  On the macroscopic scale, space is divided into two distinct regions. One region encompasses this universe, and the other encompasses the accompanying universe. These two regions are connected through countless number of microscopic passage ways through which aether is allowed to flow from this universe where it is at a higher pressure into the other one where it is at a lower pressure.

5- All spatial dimensions are straight lines, and are extending as aether is literally continually stretching the boundaries of space.

The topographical presentations of the gravitational forces near massive stars and even galaxies give the impression that the spatial dimensions are curved, while they are not. The curvature that is detected is not representative of the spatial dimensions themselves but rather indicate the gradients that exist in the flow speeds of the local aether that is headed towards the local star, the galaxy or a black hole.

Stars and galaxies (on their own scales) are dense matter concentrations in relatively small or limited regions of space. Denser matter concentrations mean more available drain holes for aether to flow through, in a given volume of space. This effect automatically encourages higher flow rates of aether towards that region of space. As it is explained in details in the section on "What is Gravity?", the speed of the aether flow is also dependent on the distance to the matter particles (or their aggregates, such as stars or even galaxies).

Particularly, the drawing representing the flow speeds of aether in the vicinity of black holes (shown below) clearly demonstrates a funnel shape with an increasing slope (faster aether flow speeds) at shorter distances from the black holes' event horizons.

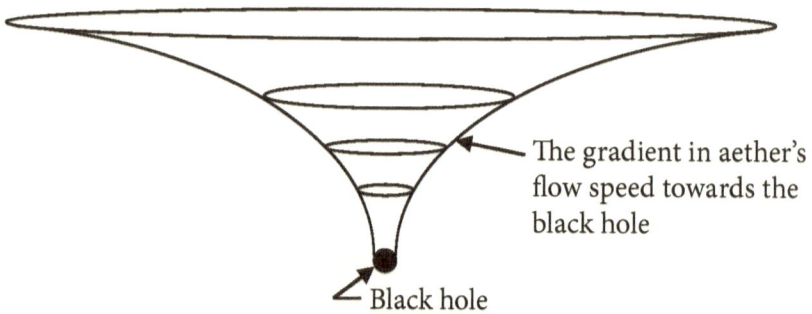

The gradient in aether's flow speed towards the black hole

Black hole

The flow of aether is from all directions

Note that, the gradual increase in the slope of the funnel shaped surface towards its center, does not represent the curvature of space, but rather the gradient that exists in the ever increasing speed of the aether as it approaches the black hole's event horizon. The very same is true about the gravitational influence of other celestial bodies such as stars and galaxies, as a whole.

6- Space, hosts this universe and its accompanying universe.

Aether medium is a continuous medium that not only houses the contents of this universe, but also its accompanying universe. The accompanying universe is not a duplicate of this universe in any shape of form. It has its own specific duties to perform. Its existence actually enables this universe to function properly. This universe and its accompanying universe are in fact complementary parts of each other.

The two universes are connected through the matter (and anti-matter) particles, as well as black holes which act as drain holes for aether to flow from this universe into the other one.

7- The portion of space that hosts this universe is larger than the portion that is hosting the accompanying universe.

This is simply due to the intrinsic properties of space that allow parallel growth of the portion that is hosting the accompanying universe so that it can be in direct contact with each and every matter particle that exists in this universe.

However, since aether is at a much higher pressure in this universe, the limits of this universe have been and still are extending faster than the limits of the accompanying universe.

Note that, the aether outer limits or frontiers in this universe do not host any matter or anti-matter particles, let alone any black holes. Therefore, the accompanying universe did not need to keep up with the expansion of this universe.

What is the Essence of Time?

What is the Essence of Time?

In physics, time can be defined as the separation of events. For example, having breakfast and having lunch can take place four hours apart. Or, two consecutive sun rises are said to be 24 hours apart. Definitions provided for units of time, are all interdependent, and none of them offers any insight into what 'time' really is. Currently the only definition that one can offer for time is that,

"Time is the gap between events."

<u>But, What really is 'Time'?</u>

So far mankind has identified and even proven some of the characteristics of time. For example:

- Even though the passage of time may vary, but it is always towards the future.

- According to Mr. Lorentz's Relativity equations, the passage of time experienced directly depends on the speed of the object in the physical space. As the speed of the object approaches that of light waves in vacuum, the passage of time will be experienced at a slower and slower pace.

- The passage of time experienced anywhere in this universe also depends on the strength of the gravitational force at that particular location. As the strength of the gravitational force increases, the passage of time is correspondingly experienced at a slower pace.

The universe has existed for a long time in the past. It does exist at the present time. And it will exist for some time into the future. Therefore, to reach a correct understanding about the structure of the universe, one must have a proper and concise understanding of time itself.

<u>"'Time' is one of the properties of the aether medium."</u>

The whole volume of the physical space is occupied by aether. The existence of aether in the physical space is exactly like the existence of the air in the earth's atmosphere, with only one difference and that is, the aether covers the whole of the physical space, not just the space between the planets and stars. The nineteenth and the early twentieth century physicists were aware of this aether and had formulated and based their respective theories on its existence.

For example, Mr. Maxwell's electromagnetism equations were based on the existence of aether in the physical space. Using the very same equations, without the need for any experimental apparatus, Mr. Maxwell was able to calculate the exact speed of light in a vacuum.

One can also name Mr. Lorentz, whose relativity equations on the relation between time and space were derived based on the existence of the aether in the physical space. The theories and equations provided by both Mr. Maxwell and Mr. Lorentz are still valid and in wide spread use. In fact, they are the backbones of the modern physics.

Light and all other electromagnetic waves are transmitted as phase vibrations through the medium of aether, exactly as sound waves are transmitted as phase vibrations through various mediums such as air. The speed of travel for any wave which is transmitted as phase vibrations in a given medium is independent of the speeds of the source or the receiver of the waves. This is why, no other waves (<u>not objects</u>) which are transmitted as phase vibrations in aether can spread at speeds that are faster than the speed of light in that medium.

"The passage of time experienced, by any object, directly depends on the speed of that object relative to the medium of aether."

As the speed of an object, such as a superfast spaceship, (relative to aether) approaches that of the phase vibrations in aether, the object experiences the passage of time at progressively slower paces. In case the speed of the object reaches the speed of phase vibrations in aether, the passage of time will no longer be experienced by that particular object.

Even if the object stands relatively still in one place in this vast universe and somehow the aether is encouraged to pass by, the resulting effect will be the very same and identical. That is, as the induced speed in the aether is increased, relative to the object in mind, the passage of time experienced by that object will be at slower and slower paces. Again, as the speed of aether relative to the object reaches the speed of the phase vibrations in aether, the object will stop experiencing the passage of time, altogether.

Based on the above information, for time to exist, the medium of aether has to exist and it must host phase vibrations. In other words,

"'Time' started its existence as the very first phase vibrations were generated in the medium of aether."

However,

"The meaningful startup point for experiencing time was when matter and anti-matter particles were formed, since as they fell behind as compared to the ripples that had formed them, they were the very first entities that experienced the passage of time."

Now, it is important to identify various natural phenomena or artificial methods, which may result in the desired effect, namely encouraging the relative motion between a given object and aether to at least approach the speed of the phase vibrations in aether. The following are a few of such natural phenomena or artificial methods:

- **The speed of the object in the physical space**
  As it was stated earlier, any increase in the speed of an object, in any direction, is an increase in its speed relative to the local aether. Therefore, the passage of time experienced by the object will be affected, according to the equations provided by Mr. Lorentz.

- **The strength of the local gravitational force field**
  According to the information presented in a separate section of this book, "What is Gravity?", the force of gravity is in fact the drag force induced by the flow of aether towards the matter particles. The speed of aether directly depends on the amount and density of matter particles that are present in any particular direction.
  Therefore, the more matter particles are concentrated in a smaller volume of space, the greater will be the speed of aether flowing towards that locality. The fastest aether flow takes place close to black holes. In fact aether's speed literally reaches the speed of light (the speed of phase vibrations in aether) at the event horizon of black holes.

- **The strength of the local magnetic field**
  According to the information presented in a separate section of this

book, "What is Magnetic Field?", the magnetic field is in fact a type of round trip motion that is induced in aether. The speed, at which aether is encouraged to move, directly depends on the strength of the magnetic field present.

Therefore, a stronger local magnetic field encourages the local aether to move faster which in turn causes a more pronounced desired effect on the rate at which time is experienced.

- **The strength of the local electric field**
  The electric field is in fact a type of round trip motion that is induced in aether. The speed, at which aether is encouraged to move, directly depends on the strength of the electric field present.

  Therefore, a stronger local electric field encourages the local aether to move faster which in turn causes a more pronounced desired effect on the rate at which time is experienced.

It should be noted that, the effects associated with the movement of the object, the local gravitational force field, the local magnetic field and the local electric field are accumulative. That is, if the object is moving in the very same direction as the aether is encouraged to flow due to strong gravitational and/or magnetic and/or electric fields, they will counteract each others' effects, since the object and the aether will be moving in the same direction. Therefore, their relative motion is actually reduced.

However, if the object is moving in the opposite direction as the induced flow in aether, the effects will be complementary to each other.

It must be emphasized that,

At nowhere in this vast physical universe aether is at a complete rest. Aether is always in motion. The continuous motion of aether in the physical space may be compared with the continuous motion of the air molecules in the atmosphere or the continuous motion of the water molecules in the oceans.

Time experienced by any object (or a living being) is directly dependent on the motion of that object relative to its local aether medium. As either the object speeds up or the aether is made to pass by faster, the object will experience time at progressively slower paces. And, if their relative speed reaches the speed of phase vibrations in that medium, the object can literally catch up with the phase vibrations that it generates.

At this point, the object will no longer experience the passage of time, because it would not distinguish between present and past. Even though it sounds strange, one could say that,

"The real time is actually progressing at the speed of phase vibrations in the medium of aether."

Therefore,

"Any object that moves at the same speed as the speed of the phase vibrations in the medium of aether, it will automatically be in sync with its own past/present, concurrently. Therefore, it cannot experience the passage of time."

Imagine a supersonic airplane flying at the speed of sound in air. As the airplane is travelling at exactly the speed of sound, using a very strong speaker, the pilot can broadcast his own voice counting 1, 2, 3, …., 100,…..

Since the airplane is moving at the speed of sound in the air, the portion of the sound waves, corresponding to each and every number stated, that are propagating towards the front of the airplane will be literally superimposed on each other, as if they were/are all stated at the very same time.

In other words, all of the numbers that were stated up to and including the one stated at the present will be heard simultaneously. Basically, the "past" starting from when the airplane reached and stabilized its speed at exactly the speed of sound in the air, and the "present" will be experienced at the very same time.

The figure below demonstrates the sound waves generated during such a scenario.

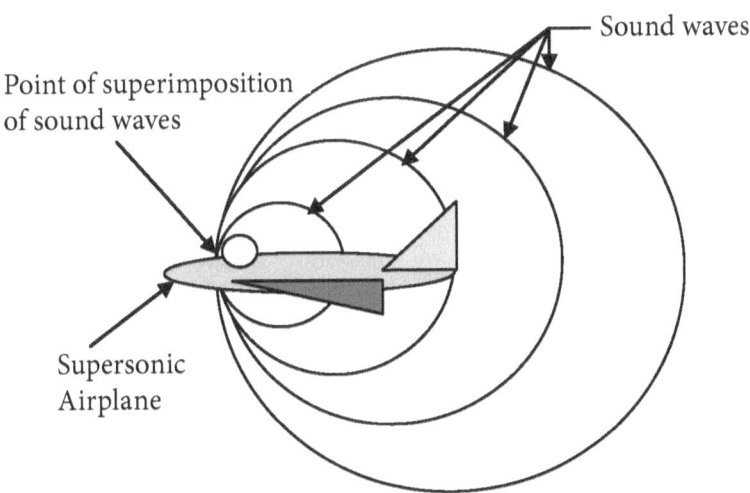

Point of superimposition of sound waves

Sound waves

Supersonic Airplane

If the pilot decides to fly his airplane faster than the speed of sound in the local air, then the sound generated due to numbers stated in the past will literally fall behind and only the number that is currently stated will be heard.

The figure below demonstrates the sound waves generated during such a scenario.

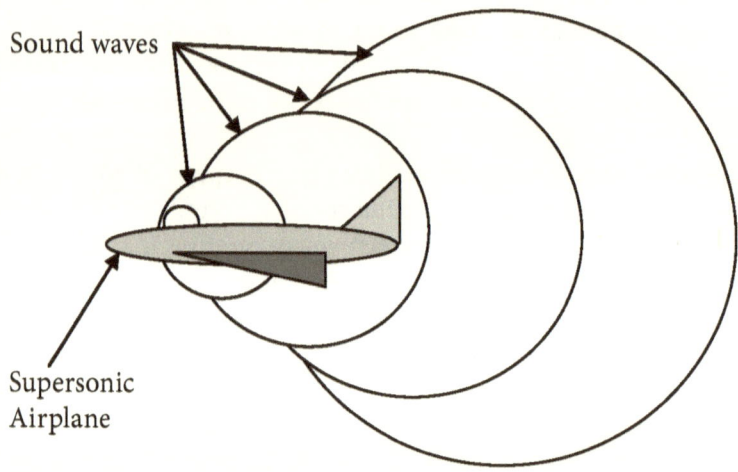

Note that, even though the airplane is moving faster than the speed of sound in air, it does not imply that the pilot will hear the sounds corresponding to any of the numbers that are not stated, yet. He will hear those numbers, one by one, as they will be stated, in their own proper timing. In other words,

**"The 'future' can only be experienced,
as it arrives and becomes the 'present'."**

Correspondingly, if the pilot decides to slow his airplane speed down to less than the speed of sound in the local air, as he is continuing with his counting exercise, the airplane will be falling behind from the wave fronts generated by the sounds.

Therefore, he will be literally falling behind from the real present that he could be at. The figure below demonstrates the sound waves generated during such a scenario.

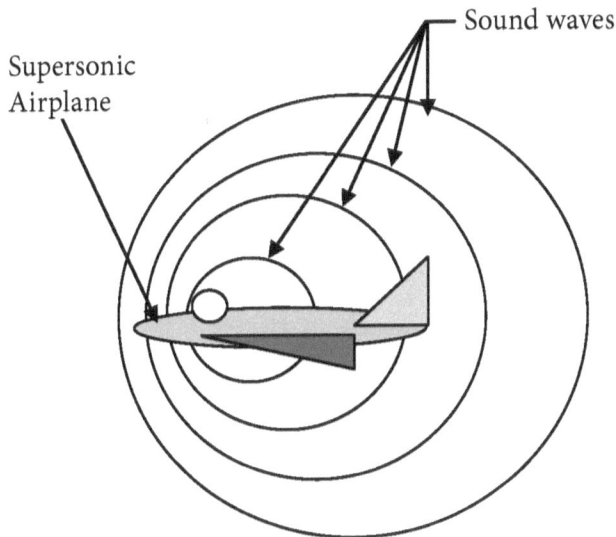

In cases where the object is relatively stationary and the local aether is made to pass by, either due to gravity, magnetic or electric phenomena, it would be as if the airplane is mounted inside a wind tunnel and the medium of air is made to pass by at various speeds. The effects experienced due to increasing the speed of the wind that is generated by the wind tunnel up to and beyond the speed of sound will be exactly the same as they were stated previously.

In short, one can say, any object that is moving at a speed that is slower than the speed of phase vibrations in aether, is literally falling behind from the wave fronts associated with its past. Therefore, in normal circumstances,

"Objects (including living beings) experience their respective delays in time and not the passage of time itself."

In other words,

"Objects and living beings are in fact experiencing time in reverse, as they are literally falling behind from the real current time (Present) that they could and should be at."

What is Light?

What is Light?

Light makes the world visible. Yet, not much is really known about its true essence. Over the centuries, man has come up with a variety of ways to describe how light behaves in different situations, but has not really paid any attention to its essence.

Particularly during the 19th century, scientists believed that light was a type of wave, namely electromagnetic wave, which is in fact a phase vibration in a medium that they called aether, since they had noticed that light behaves exactly like sound in air.

In 1865, Mr. Maxwell proposed his theory on electromagnetism. Using his formula he quite accurately calculated the speed of light in vacuum. Mr. Maxwell had based his theory on the existence of some kind of medium through which light and other electromagnetic waves were propagating. This medium at that time was referred to as aether. What aether was made of, or what kind of properties it had was not known. It was necessary to be there as a medium for electromagnetic waves to travel in. The aether medium to electromagnetic waves was thought to be just as air is to sound waves.

Over time the scientists became more and more curious about the aether and tried to understand its behavior. They thought if aether was stationary in space and all waves and even planets were passing through it, then one should be able to detect the drift effect on the surface of earth. Knowing that the earth is orbiting the sun at about 107,000 Kilometers per hour, various attempts were made to detect this drift velocity.

The most promising of all the proposed experiments were designed and performed by Mr. Michelson and Mr. Morley in 1887. Again, these two gentlemen had assumed that aether was stationary in space and earth was moving through it. They had designed their equipment with sufficient accuracy to detect such a minute velocity variation caused by earth's motion around the sun as compared to the speed of light in a vacuum.

Their experiment was based on the comparison of wave patterns generated by the same monochromatic light passing through two paths, one in the direction of the earth's motion around the sun and the other at 90 degrees to the said direction. By reflecting back and superimposing these two light branches, Mr. Michelson and Mr. Morley were hoping to see the formation of some sort of interference between the two.

They tried their experiment at various locations on the surface of the earth. They also conducted their experiment up at high altitudes, with the help of a balloon. However, every time the result was the very same. No interference pattern was formed in any of their attempts. Of course, some scientists offered suggestions indicating that the planet earth could be carrying a thin layer of this aether medium with it, as it is going around the sun. This could explain why no effect was detected.

In 1904 two scientists, Mr. Fitzgerald and Mr. Lorentz, independently, provided reasoning for detecting no trace of any drift effect in all of the experiments. Mathematically, they proved that the length of the experimental apparatus's arm which was in the direction of earth's motion would have been affected by the motion itself and therefore had resulted in annulling the intended effect of the aether drift.

Mr. Lorentz, in 1904, proposed a theory based on which the effect of motion on the apparatus was clearly explained. This theory was the first theory ever to talk of effects observed due to motion at relatively high velocities. Particularly, the effects such as the shortening of the length of the object in the direction of motion, as well as the slowing down of the passage of time, as one approaches the speed of light in a vacuum, were first proved by Mr. Lorentz' theory and his transformation equations. These transformation equations are also known as Mr. Lorentz' Relativity Theory or "Principles of Relativity".

Again, it should be noted that, Mr. Lorentz and Mr. Fitzgerald had proposed their theories without rejecting the existence of the aether medium, which was commonly accepted by the then-current scientific community.

In 1905, Mr. Einstein used Mr. Lorentz' "Principles of Relativity" and added the following two assumptions:

1- Mr. Einstein rejected the existence of the aether altogether, and claimed that light is made of particles, namely photons.

2- Mr. Einstein claimed that the speed of light in a vacuum is the maximum possible speed and it is a constant which is independent of the observer's state of motion relative to the source.

Note that:
1- The experiments performed by Mr. Michelson and Mr. Morley, by no means had indicated that there was no aether in space, or that the speed of light was the maximum speed possible.

A revised version of Mr. Michelson and Mr. Morley experiment is proposed in a separate section of this book.

2- Since Mr. Zwicky, a Swiss astrophysicist, published his report on gravitational effects due to some kind of invisible mass (matter) in 1933, some new terms have been added to astrophysicists and physicists' vocabulary, namely the dark matter and the dark energy. Dark matter apparently does not directly interact with known forms of matter. Yet, its gravitational effects on known Matter have been observed since Mr. Zwicky's time.

The relative amounts of dark matter, dark energy and regular matter/energy in this universe are given below:

73%, Dark energy,
23%, Dark matter, and
4%,   All known forms of energy AND all known forms of visible matter.

Since Einstein, some scientists have accepted that light is made of particles, namely photons which by definition have no mass. In other words, a definition that in itself is contradictory. Photons are supposedly some kind of **particles** (a real physical entity) that supposedly actually exist, yet as they get stopped, there is nothing there. That is, no mass is gained by the absorbing target, regardless of how intense the light beam may be or for how long the absorption process may last. Although, the target may heat up indicating the absorption of the energy received.

However, the majority of the present-day scientists believe that light is both, particle and wave. Their view is due to trying to quantize the light phenomenon and fit it in quantum theory equations. They can explain many of light's behaviors, as they call it a wave when it suits better, or call it and treat it as particles when it suits better, in different circumstances. **Basically, physicists are not sure what light really is.**

In order to explain what light really is, one needs to propose a theory that can provide consistent explanations for all of the known and proven characteristics of light. Also, if possible, it should predict specific verifiable new findings about light and its characteristics.

A new theory of Light

"Light is a form of phase vibration in the medium of aether."

Light and other electromagnetic waves travel as phase vibrations through aether, just as sound waves travel in air.

The aether that is referred to in this presentation is the very same aether that was believed to exist by the 19th century scientists. However, unlike it was assumed earlier, aether is not a stationary medium in space. It is a constantly moving and dynamic environment.

Aether is also responsible for various physical and energetic phenomena such as the force of gravity, magnetic field and electric field. As an example, the force of gravity is due to the drag force induced by the aether flow towards individual subatomic particles. The flow of aether drags the other particles, as well as waves that are generated in it, just as air drags a sailboat or sound waves.

Different types of motions that are constantly in progress in the aether medium are just like the various motions of air molecules that make up the atmosphere or water molecules that makeup the oceans.

The following are various behaviors and characteristics of light, as they are explained according to this new theory.

1- Light, being a phase vibration in aether can only propagate where there is aether, which is literally everywhere in this universe.

2- The **actual speed** of light in aether is only dependent on the local aether density, regardless of any and all types of motions that may exist in the local aether medium. Higher aether densities result in slower propagation speeds by light and other electromagnetic waves.

3- The **actual speed** of light in the local aether is always the same and equivalent to the speed of phase vibrations in that medium, regardless of the medium being void of matter particles and other phenomena such as gravitational, magnetic or electric fields, or not.

4- The **apparent speed** of light, between its source and its receiver, is the fastest when there are literally no deflections in its path. In other words, light follows a straight path when there are no cross flows of any kind in the aether medium. This is the case only at places far away from galaxies.

5- The **apparent speed** of light, between its source and its receiver, is reduced if the medium of aether is occupied by matter particles. The presence of matter particles, in the immediate vicinity, automatically induces a variety of cross flows in the local aether, causing light to follow a zigzag path.

6- The **apparent speed** of light, between its source and its receiver, in any medium such as a gas, a liquid or a solid, which is in fact a medium that is literally superimposed on the aether medium, is directly dependent on:

- How densely the matter particles are packed.
- How the matter particles are arranged / organized.
- Whether the matter particles are rigid in their respective positions, as in a crystalline lattice structure, or they are free to move around, as in a gas or a liquid.

7- As light is encouraged to follow a path through a crystalline structure, its **apparent speed** through that matter medium will depend on:

- The angle at which it enters that crystalline matter medium, since the number density of matter particles in any given direct is different and it is fixed.

- The frequency of the light wave, since once the light waves enter a given crystalline matter medium, they will encounter matter particles in specified distances that will be integer multiples of certain wave lengths (corresponding to certain frequencies).
  Therefore, those frequencies will be deflected / distracted more often than the other frequencies, which means they will require longer time to pass through the matter medium.

8- The coherency of the light wave passing through any matter medium directly depends on the organization, as well as the uniformity of the matter particles that make up that medium. For example, a piece of glass that is made of a single and uniform crystalline structure will allow the passage of light waves without causing much of dispersion.

However, if a piece of glass were made of many crystalline pieces that were put together or it consisted of many inhomogeneous regions, due to the manufacturing process used, every time the light waves enter a different crystalline region, they will get dispersed in different directions. This is why objects seen through some glass (or other crystalline) materials can be quite clear, while others may lead to the generation of duplicates of the image received on the other side, or may even barely allow the passage of light and only show a blurred view of the object.

9- Light, being a phase vibration, does not get absorbed by the matter particles, as it encounters them. Only the aether is absorbed due to flowing right into the matter particles. However, as light follows its path in the aether medium which is flowing towards various particles (near and far), it does get dragged/deflected temporarily as it passes through. Therefore, instead of following a straight path, it follows a zigzag sort of a path.

If the matter particles are too densely populated and are not organized in say a crystalline lattice structure, they will induce complicated deflections in the path of light waves so that as light waves pass through that medium they will spread in all random directions, and get mixed up to the point that they will weaken and eventually become totally unrecognizable.

10- Always some (even if it is reduced to a miniscule amount) light (or electromagnetic waves, in general) crosses a matter medium, regardless of its internal structure and/or thickness.

11- The vibrations due to any light wave continue to spread and weaken, yet there will be a reminiscence of them in this universe, just as the sound waves generated in the atmosphere anywhere on the planet spread in all directions and can eventually reach even the opposite side of the planet.

The only way to truly stop the passage of light waves, through a given matter medium, would be by literally blocking the flow of aether through that particular matter medium. This is an impossible task to accomplish, since aether can and does flow even through huge stars, let alone smaller objects.

Light waves generated in and carried by the aether medium can even be likened to the waves that are generated in a river stream which along its path encounters many rocks and stones of different sizes (among other objects and directional detours), and eventually the original wave pattern literally disperses into countless other waves. The waves are still there but they will be so weak that their detection is only dependent on the sensitivity of the equipment used.

It is due to their sensitive detection (reception) abilities that humpback whales can communicate over distances that are in excess of 1,000 km.

Note that: If a sensitive enough a receiver is built, one should be able to not only hear the sounds that were generated, but also literally witness events that have occurred years ago. The sound waves and light waves of the past are literally still floating in the atmosphere and space, respectively.

Application of the theory
Providing explanations for known light behaviors

A variety of characteristics of light that are known and proven by various international scientific groups are listed below:

1- As light passes close enough to a galaxy (or even a star), its path bends towards the galaxy (or the star). In other words, the gravitational pull of the galaxy as a whole (or star) affects the path of light as it passes by.

2- If light crosses the event horizon of a black hole, it will be pulled into the black hole.

3- The farther the galaxies are from the earth, the faster they are getting away from it. The rate at which a given galaxy is receding from the earth is calculated using the amount of red shift that is detected in the light spectrum (frequencies) received from that particular galaxy.

According to these findings:

1- Gravity can change the direction in which light is propagating.

2- Gravity can literally stop the motion of light.

3- Relative motions of the source and the observer only affect the frequency of the light received.

4- According to astrophysicists and physicists, the speed of light in vacuum is constant and independent of the relative velocities of its source and its receiver.

When the above effects are considered, one can see that they are quite similar to the way sound is affected as it spreads through air. For example, using two parabolic dishes, positioned so that the sound generated at the focal point of one can be heard at the focal point of the other, one can perform the following experiments:

1- The effect of a gentle cross wind on sound waves traveling in the air is the same as the effect associated with the existence of a star (or a galaxy) near the path of a light beam. In this case, as it is shown in the figure below, the breeze literally drags/carries the sound waves downstream. Therefore, to be able to hear the transmitted sound wave, one needs to point both the transmitting and the receiving dishes in the up-stream direction. The required amount of adjustment will depend on both the distance between the two dishes and the speed and direction at which the breeze is flowing.

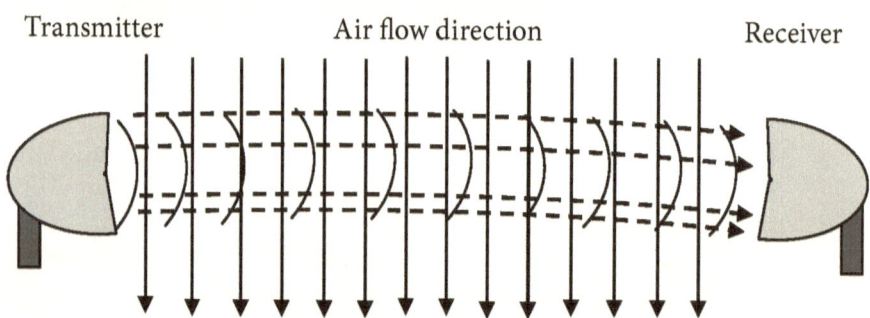

Transmitter   Air flow direction   Receiver

Likewise, as it is shown below, a light beam experiences the very same dragging effect by the aether medium that is flowing towards a star or a galaxy that is near its path.

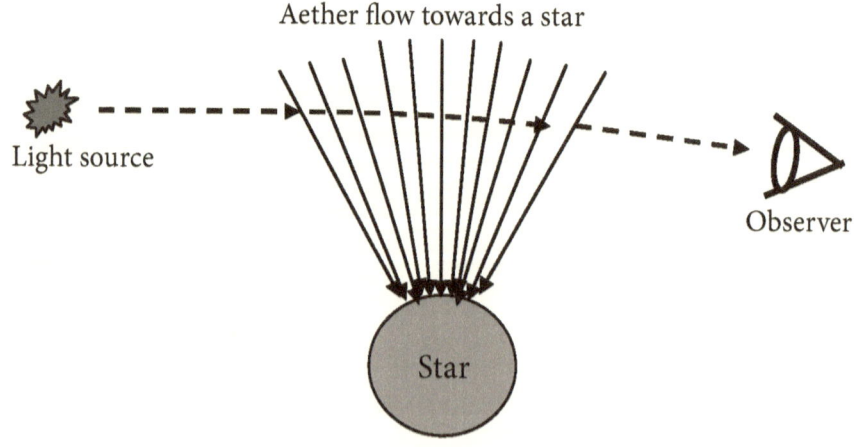

Aether flow towards a star

Light source

Observer

Star

2- When the wind is blowing at or faster than the speed of sound and it is in the direction that is in-line with the two dishes, from the receiver towards the source of the sound, the transmitted sound wave would never reach the receiver. This effect is analogous to the existence of a black hole in the immediate vicinity of the source of light, absorbing any light that crosses its event horizon.

Another way of describing the effect of a black hole using sound waves is the case of two supersonic airplanes flying in line, one in front of the other. The pilot in the front airplane, no matter what kind of or how sophisticated an equipment he uses, he will not be able to hear the sound generated by the engines of the airplane following him. This is due to the fact that, the sound waves in the open medium of air is literally falling behind.

One may also use a supersonic wind tunnel to perform this test. In this case, he needs to install a speaker (connected to a stereo system) near the outlet and a microphone (connected to a set of head phones) near the inlet of the supersonic portion of the wind tunnel. The experimental setup is shown below.

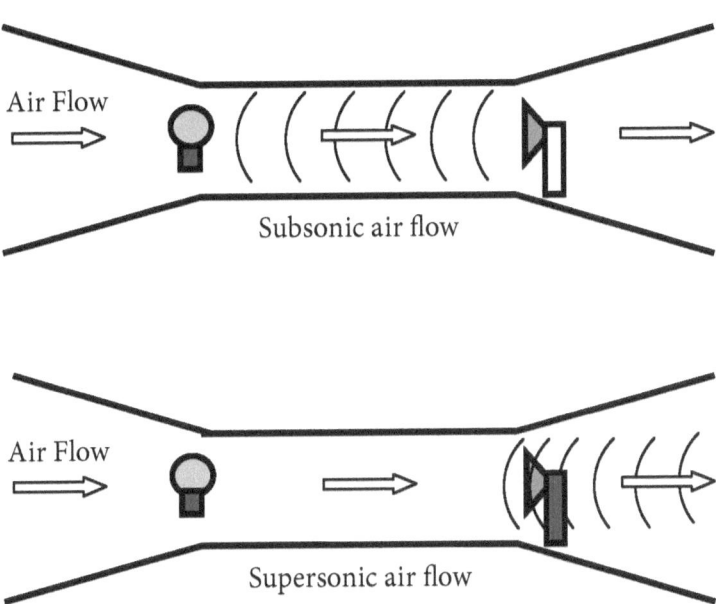

By starting the wind tunnel and bringing it up to its supersonic speed, while listening to his favorite music; he/she will notice that the sound becomes more and more low pitched. And, eventually as the sonic barrier is crossed, he/she will not hear the music at all. This type of test clearly demonstrates the effects that the motion of the medium has on the transmitted phase vibrations.

This is exactly what is taking place in the immediate vicinity of a black hole. A black hole is made of an enormous amount of particles (drain holes) gathered together in a very small volume of space. For more detailed information in this regard, please refer to the sections on "What is Gravity?" and "Black Holes and Their Properties".

The induced inward flow of aether at close proximity of any black hole is so great that the aether is moving faster than the phase vibrations such as light or other kinds of electromagnetic waves in that medium. Therefore, light simply cannot propagate up-stream fast enough to counter the speed at which aether itself is flowing in the opposite direction and into the black hole, and hence the name black hole.

3- The sound waves from any source to any receiver are transferred at exactly the very same speed via the phase vibrations of the atoms and molecules in a medium such as air. This speed is independent of the relative motion of the source and the receiver. It only depends on:

- The composition of the local medium, say air, as in types and percentages of different atoms and molecules that are present,
- The local temperature and pressure, and
- The motion of the medium itself (as a whole) relative to both the source and the receiver.

In other words, the observed behavior of light in a vacuum and particularly its being stopped by the gravitational pull of the black holes are clear indications that, light is phase vibration in aether (a wave), as it was already known to be by the 19th century scientists, and its speed and its direction of propagation are affected by the motion of aether.

Other known facts about light

The following are some of the other known facts about light and how it behaves in different circumstances.

1- Light demonstrates different apparent speeds as it travels through different mediums.

Light and other electromagnetic waves, travel the fastest in outer space and their speed in various materials is dependent on:

- The density of the medium.
- The frequency of the light itself.
- The angle of incidence, as it enters the medium.

As it is explained in details in the section on "What is Gravity?", matter particles act as drain holes for aether. Therefore, their existence automatically encourages the flow of aether towards them, forming complex cross flow patterns in the aether medium.

As light waves (carried as phase vibrations) pass through, they get dragged by these local cross flows induced in aether and mandatorily follow a zigzag pattern. The more matter particles present near/along the path of a light beam, the more detours its waves will be encouraged to follow; hence it will take longer time to cross through the medium.

The dependence of the speed of light in a given matter medium on the light's frequency and/or angle of incidence, particularly in crystalline materials, are due to the number density and spacing between the matter particles that are encountered in any given direction, as light enters the matter medium. Uniform crystalline structures result in more dependence of the speed of light on its frequency. Again, the more compact the matter particles are organized in the matter medium in any given direction as compared to the others, the slower will be the speed of light in that particular direction, since it will get dragged more frequently and hence it will have to travel longer actual distances (in a zigzag pattern) within the matter medium.

2- Light is transferred through some materials, while it gets partially or even totally absorbed in others

If the matter particles (atoms or even molecules) are organized in a crystalline structure, all through the thickness of the medium (as in a single piece of perfectly homogenous glass), then the path is more or less clear with symmetric zigzag detours, detours that on average cancel each other in how much they cause the light waves to be dragged/deflected towards one side and then towards the other. Therefore light waves follow a general straight path as they go through the medium. (See drawing below)

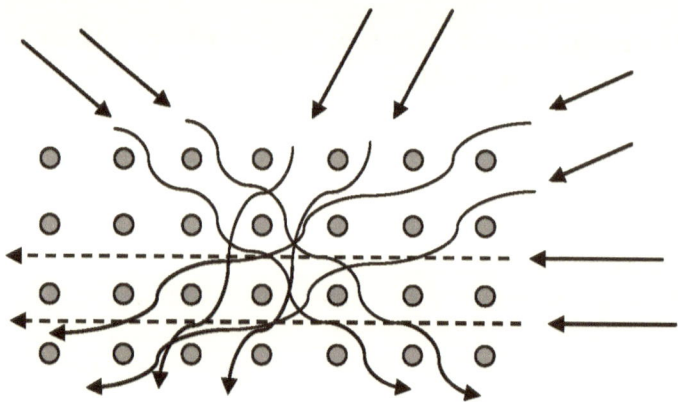

An example of a crystalline material

The same is also almost true when the medium consists of particles (atoms and/or molecules) that are in total random motion, as in gasses or liquids. In these cases, the light wave gets deflected in random as well, but it is allowed to continue, particularly when the particle density of the medium is not high enough to cause the total dispersion of the incoming light wave before it gets the chance to reach the other side of the matter medium. (See drawing below)

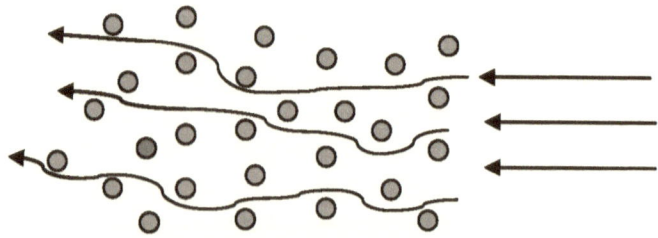

An example of a gaseous or liquid type of medium

However, if the matter particles in the medium are not organized but they are densely packed and are rigid in their respective locations, then the light waves get literally stranded by becoming dragged/deflected in complicated patterns, by the cross flows that are induced in aether, as aether is flowing towards individual matter particles. Denser and more unorganized matter mediums lead to more severe zigzag patterns in the path of light waves passing through, hence causing them to demonstrate a slower apparent speed. (See drawing below)

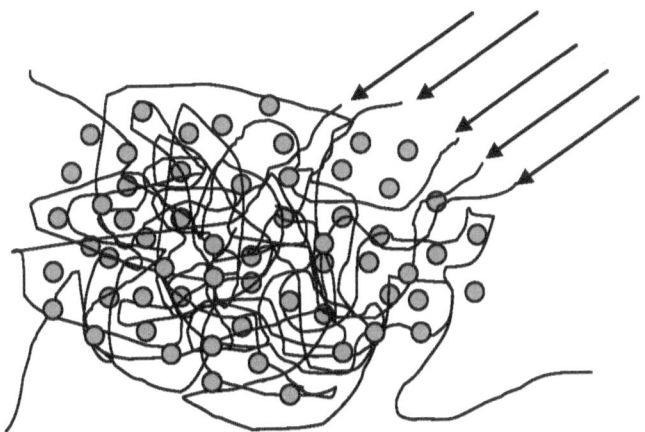

An example of an opaque material

Note that, always some light waves (even if it is reduced to miniscule amounts) cross a matter medium, regardless of its internal structure and/or thickness. However, the thicker or more opaque materials require lights that are of higher intensity.

3- The photoelectric effect

Light waves that are of certain frequency or higher can excite electrons, in certain atoms, and encourage them to let go of their atomic bonds and freely float to other atoms. This process is referred to as the photoelectric effect. The photoelectric effect can best be explained as follows:

The waves that are formed on the surface of water in a pond demonstrate very gentle vertical movements. The waves can have different amplitudes but still their vertical motions are always quite gentle and slow. This slow and gentle wave motion in water is mainly due to the high density of water medium.

Any object floating on the water, such as a piece of paper, wood or even a small boat, will also demonstrate the very same alternating up and down motions. Of course, each will move by different amounts due to their own overall weights as well as their respective sizes as compared to the overall shape and size of the wave (ripples) encountered.

Now consider a medium such as air where the sound waves can be quite packed closely and can have fairly high amplitudes. In these cases, the vibrations will cause the dust particles or even small objects resting on the surfaces that are in the way, to vibrate violently.

If the amplitude is fairly high and the frequency is fairly low, as in low base sound at high volume, it will promote abrupt motions in the tiny objects encountered. Those objects will in turn manifest their responses by separating from the surfaces they are resting on.

If the frequency is high enough and the amplitude is also fairly high, the vibrations will cause serious internal effects in the surrounding materials, such as breaking glasses and so on.

Regarding light and the photoelectric effect that it can induce in certain materials, one can visualize the waves that are propagating in aether causing the very same type of effects on the electrons, just as were described to take place due to waves in water and air.

In this case, electrons, due to their being quite a bit lighter than the other subatomic particles (by a factor of about 1,840 to 1), are readily influenced by light waves. The light waves being phase vibrations in aether, literally cause the electrons to bounce in different directions as they encounter them.

If the frequency of the light wave encountered is low the electrons will simply follow a roller coaster kind of a motion, just as a small boat does when it encounters the waves generated by a ship passing by. In this case, the electrons do not gain sufficient momentum to break free from their atomic bonds, hence they just simply wobble as they follow their orbital paths around the nucleus.

However, if the frequency of the incoming light wave is above a certain minimum, the light wave will cause the electrons to be literally shaken out of their stable orbital paths and break free of their respective nuclei.

This effect is exactly like shaking dusts or debris off of a carpet by violently shaking the carpet so that its surface simulates a wave motion. If the generated wave motion (the ripple) is abrupt and strong enough the particles of dust and debris will be shaken off, otherwise they will hold on to their respective positions and just enjoy the ride.

Once a light wave having above certain minimum frequency (which corresponds to having more than a certain minimum amount of energy carried by the wave form) encounters certain atoms, the electrons in the outer layer have no choice but to let go of their bonds and start to wander between atoms. The light intensity will only affect the number of electrons that can be freed.

Note that, if the frequency of the incoming light is not high enough, the electrons will not let go of their bonds with their respective nuclei, regardless of the intensity of the light beam used. They would only experience more wobbles, but will still manage to hold on to their respective nuclei.

4- Reflection, Refraction and Interference of light waves.

These properties of waves, particularly relating to light waves, are well explained, documented and understood by all.

Conclusions

Light and other electromagnetic waves are phase vibrations in aether, just as sound is in a medium such as air. They do not consist of some kind of particles.

In the absence of any matter particles, the apparent speed of light is equal to its actual speed in aether. In these cases, the light waves travel in a straight line.

However, in any other scenario where there are matter particles present, particularly when they are nearby, due to local cross flows of aether induced towards individual matter particles, the path followed by the light waves becomes zigzagged. Denser matter mediums give rise to more frequent drags and deflections that are mandatorily experienced by the light waves, as they pass through. Therefore, light will require more time to cross the medium, since it follows a path that has become ever more lengthened by the more rigorously zigzagged patterns that are formed due to cross flows in the local aether medium.

The phase vibrations (such as light and other electromagnetic waves) in aether are also shown to be responsible for the photoelectric effect, given they posses above certain minimum energy/ frequency to literally shake the electrons off of their orbital paths.

What is Gravity?

What is Gravity?

The force of gravity is one of the main forces in nature. It is also the most common force that man has to deal with in his daily life. That is why, particularly over the last 400 years, man has developed solid and reliable theories regarding the laws that govern its behavior.

Mr. Galileo proved that all objects regardless of their density fall towards the earth at exactly the same rate. Mr. Keppler formulated the laws governing the motions of the planets around the sun. Mr. Newton, with the help of an apple, provided the basic laws of gravity which apply not only to all objects on this planet, but also to all of the contents of this universe.

Using Newton's theories and formulas, man has gained remarkable amount of knowledge about the structure of the universe. Of course, especially over the last century, there have also been such theories as the general theory of relativity proposed by Mr. Einstein in 1917 that tried to relate the force of gravity to the imaginary curvature of the three dimensions of the physical space.

The famous transformation equations provided by Mr. Lorentz in 1904 (relating space and time) and the equations developed by Mr. Maxwell in 1865 (expressing his electromagnetism theory) which are still valid and in general use today were based on the existence of a medium which was referred to as aether. Aether was believed to be the medium through which light and other electromagnetic waves were transmitted.

In 1905, Mr. Einstein came up with his special theory of relativity, which was based on Mr. Lorentz's transformation equations (Principles of Relativity). However, Mr. Einstein, without providing any proofs, rejected the existence of aether and also claimed that the speed of light is the fastest in vacuum and basically nothing can even approach that speed.

Interestingly enough, since 1933, starting with Mr. Zwicky, some new terms have entered the scientific vocabulary, namely the dark matter and dark energy. Dark matter and dark energy are found to be literally everywhere. The amount of dark matter in the universe is calculated to be more than five times the amount of regular matter which all stars and planets are made of. The density of the dark matter is known to be greater towards the outside of the solar systems and galaxies. In fact, there seems to be halos of dense dark matter particularly surrounding galaxies. In other words, there is no such thing as a vacuum, since dark matter is literally everywhere.

According to various theories, black holes are unique entities in this universe which possess such a strong gravitational force that not even light or any other type of electromagnetic wave can escape their grips.

One of the basic mysteries (and the greatest) about the force of gravity is the speed at which it acts. All gravitational computations regarding the navigation of any inter-planetary probe is done using the instantaneous location of every planet and not where they look to be from the probe's position at any given point in time. Also, according to direct physical observations and data collected, at any given instant, the planet earth is pulled towards the true instantaneous location of the sun and not where it looks to be.

In other words, even though it takes about 8.3 minutes for the sun's light to reach the earth, apparently it takes no time at all for the force of gravity to travel this very same distance. This is in direct contradiction with the underlying principle of special relativity. According to Einstein's theory of relativity nothing can travel faster than the speed of light in a vacuum, while the force of gravity has been doing this all along and it will keep on doing so from now on, as well. In short,

At the present time, there are no theories that can explain the instantaneous effects of the sun's gravity on the earth as well as the other planets.

Again, what is gravity?

To explain the force of gravity and its properties, one has to first acknowledge the following known facts concerning gravity in this universe.

1- Galileo's experiments,
2- Accumulative effects of gravity,
3- The governing laws of gravity,
4- The bending of light, as it passes nearby a star or a distant galaxy,
5- The extreme force of gravity near black holes,
6- The existence of dark matter in the entire universe,
7- The density gradient observed in the dark matter concentrations in the solar systems and galaxies,
8- The existence of dense halos of dark matter surrounding galaxies,
9- The instantaneous effect of the gravitational force, with speeds that by far exceed the speed of light in space,
10- The widening of the planetary orbits around the sun.
11- No direct interaction between dark matter and regular matter,
12- The Michelson and Morley experiments, as well as,
13- The initial sudden and yet temporary expansion of the universe during its infancy.

The simplest way to explain what gravity really is and how it operates, is through providing explanations on how all of the known facts mentioned above are possible.

A New Theory

This theory is based on two assumptions:

1- The dark matter which has been confirmed to exist everywhere in this universe, since 1933, is in fact aether, and

2- Any and all sub-atomic particles in this universe are in fact tiny drain holes (not black holes, but drain holes), just like tiny tubes, through which aether which is at high pressure inside this universe flows into its accompanying universe.

Note that, the accompanying universe also has 3 dimensions and it is accessible only under certain conditions.

The accompanying universe is not a duplicate of this universe in any shape or form. It has its own duties to perform. It is a complementary part of this universe. Its existence enables this universe to function properly.

As a first step, for simplicity, consider one sub-atomic particle that is freely floating in space. The incoming flow of aether to this sub-atomic particle will be from all spatial directions. The flow will extend literally to the end of space, and it will always be directed towards the location of the subatomic particle. The flow of aether, towards this particle, is shown below (shown in a plane, due to simplicity).

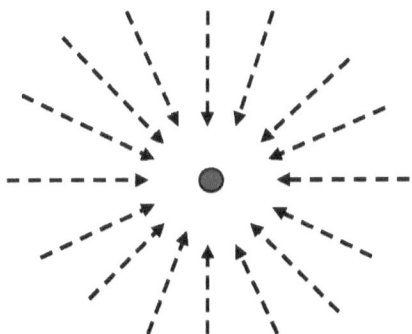

The flow of aether is directly towards the particle

As it is shown in the drawing below, any other particle would experience the effect of this induced aether flow as a drag force towards the sub-atomic particle.

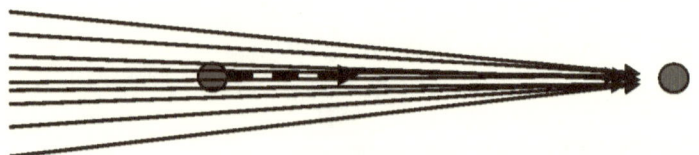

The drag force will always be towards the true location of the sub-atomic particle.

This drag force is the force of gravity that is experienced by anything and everything in this universe.

Now, let us explain each and every one of the known gravitational phenomena, using **this new theory** which states that **gravity is in fact the drag force induced by the flow of aether**.

1- Galileo's experiment

All objects are made of the very same three elementary particles, namely protons, neutrons and electrons. All of the elements are only different combinations of these three particles. And, different objects in nature are made possible by different combinations of various elements.

According to this theory, each and every particle experiences the force of gravity (such as that of the earth) independent of all other particles. Therefore, either they are as individual particles or as parts of an atom, a molecule, or an object of any size or type, they all will be pulled towards the earth at the very same rate, given there be no air resistance (collisions).

This is how Mr. Galileo was able to propose that even a feather and a piece of Iron will fall at the very same rate, if there be no air resistance.

"The constancy of the gravitational acceleration (g factor) at any given elevation is also due to the constancy of the drag force induced by the flow of aether at that particular elevation (from the center of mass of earth), regardless of the longitude and latitude."

2- Accumulative effects of gravity

Since the speed of aether rushing towards any given region of space depends only on the overall number/size of openings or drain holes available, the more the number of sub-atomic particles present the more aether is automatically allowed to go through. This is analogous to a strainer; more open holes automatically lead to more fluid flowing through the strainer as a whole.

Therefore, as many sub-atomic particles are grouped together and form a variety of atoms, molecules, objects, planets, stars or even galaxies, the overall induced aether flow towards them is cumulative and hence the force of gravity generated becomes stronger and stronger.

3- The governing laws of gravity

It is clear that the aether flowing towards any sub-atomic particle is coming from the volume surrounding the particle. Since the surface area of the sphere covering all around any particle is given by the formula, (4π), one can see that the total surface area of the sphere will change by the square of the distance (radius, r) to the particle.

Therefore, in order for the flow of aether to be consistent, the speed at which aether has to move towards the particle also has to change accordingly. For example, at half the distance the spherical surface area surrounding the particle is reduced to a quarter. Therefore, the speed of aether has to quadruple. Hence, the drag force generated by the motion of the aether will also be quadrupled.

Note that, the speed at which aether flows, changes by a bit more than the quadruple, at half the distance. This is due to minute reductions in the aether density/pressure as it approaches the sub-atomic particles or their collectives, such as planets or even stars.

4- The bending of light, as it passes nearby a star or a distant galaxy

In this case, one has to first understand the true nature of light itself. According to existing theories, light is an electromagnetic type of wave made of weightless particles, namely photons, which travel through the vacuum of space. This definition of light was introduced by Einstein through his special theory of relativity.

To come up with a better understanding of the true nature of light one has to first accept the existence of aether as the medium used by light to travel through, the very same aether that was accepted by the nineteenth and the early twentieth century scientists.

Light in aether acts exactly as sound does in air. Light and all other electromagnetic waves are in fact phase vibrations in the medium of aether which exists everywhere. Therefore, all laws governing the propagation of sound waves in air also apply to propagation of light in aether.

Just as sound waves get deflected/dragged by passing through a cross wind, so do electromagnetic waves as their associated phase vibrations pass by a star which induces relatively high speeds of aether cross flow towards its respective center. This cross flow of aether literally drags the phase vibrations associated with light and other electromagnetic waves towards the star.

The similarities between the bending in the direction of propagations of sound and light are demonstrated in the following figures.

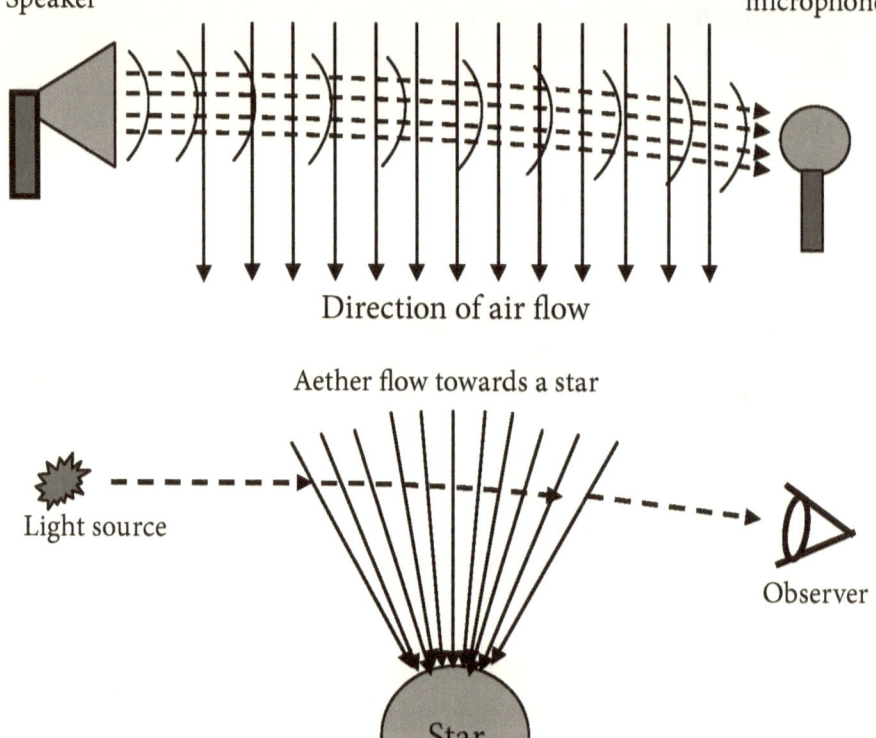

Only a section of the aether flow is shown in the figure.

5- The extreme force of gravity near black holes

Let us first consider what happens to sound propagation in air as the air itself is encouraged to move in the opposite direction at a speed greater than the speed of sound. In this case, the sound waves cannot compete with the flow of air in the opposite direction. Therefore, no sound propagation takes place in the forward direction.

One may use a supersonic wind tunnel to perform the following test. He needs to install a speaker (which is attached to a stereo system) near the outlet and a microphone (which is connected to a headphone set) near the inlet of the supersonic portion of the wind tunnel.

By starting the wind tunnel and gradually bringing it up to its supersonic speed, while listening to his favorite music; he will notice that the sound becomes more and more low pitched. And, eventually as the sonic barrier is crossed, he will not hear the music at all.

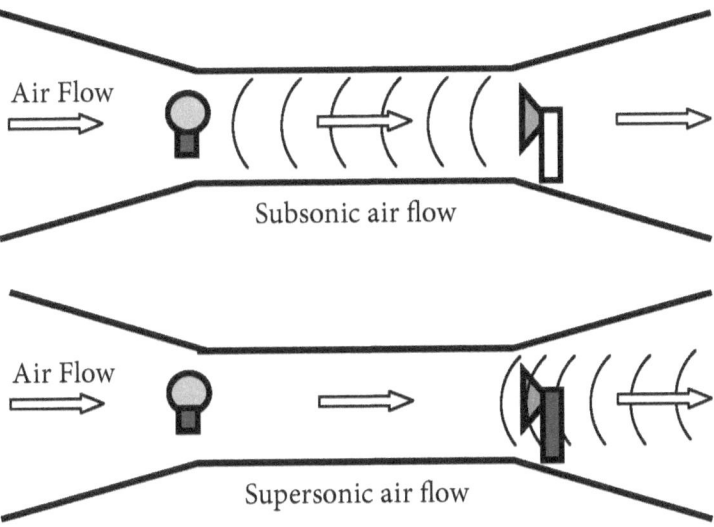

Subsonic air flow

Supersonic air flow

This type of test clearly demonstrates the effects caused by the motion of the medium on the transmitted phase vibrations. This is exactly what is taking place in the immediate vicinity of a black hole. A black hole is made of an enormous amount of matter particles (drain holes) gathered together in a very small volume of space. The induced inward flow of aether at close proximity of any black hole is so great that the aether is moving faster than phase vibrations such as light or other kinds of electromagnetic waves in that medium.

Therefore, light simply cannot travel up-stream fast enough to counter the speed at which aether itself is flowing in the opposite direction and into the black hole, and hence it gets dragged into the black hole.

6- **The existence of the dark matter in the entire universe,**

7- **The density gradient observed in the dark matter concentrations in the solar systems and galaxies, and**

8- **The existence of dense halos of dark matter surrounding galaxies**

The flow of aether into sub-atomic particles is pressure driven, meaning the aether in this universe is at a much higher pressure as compared to the aether that is in the accompanying universe. Therefore, there is a slight gradient in the density of aether over long distances. This density gradient is due to the flow of aether and also encourages its flow, as well. Even though aether's density gradient is very gradual, it becomes more pronounced over long distances and especially around larger aggregates of particles, such as stars, solar systems and particularly galaxies.

Note that, the slopes of the induced gradients are greater at shorter distances and gradually decrease to their minimum values with distance. For this very reason, it is expected to find the outer regions of the solar systems and especially galaxies to be hosts to thicker and denser concentrations of aether.

9- **The instantaneous effect of the gravitational force**

The flow of aether and hence the force of gravity induced by any object, be it a star or be it a planet, is done independent of any and all other celestial bodies. Other bodies experience the force of gravity as they are pulled or dragged by the flow of aether present at their particular locations. This is why the earth is affected by the sun's force of gravity in a direction which is towards sun's instantaneous position.

Note that, the force of gravity generated by the sun in its surroundings has nothing to do with the force of gravity generated by the earth or any of the other planets in their respective surroundings. However, each of these heavenly bodies, namely the sun, the earth and the other planets, at their current respective positions, are instantaneously affected by the forces of gravity that are generated by the others.

The flow of aether created by and towards the sun, for instance, is from all around and it is totally independent of the existence of other celestial bodies such as planets.

The force of gravity generated by the sun is always directed towards its center. Therefore, as planets, including earth, go around the sun, they are literally affected instantaneously by the crossflow of the aether; a flow which at all times is directed towards the center of the sun. Hence, the reasoning for the detection of the gravitational force of the sun on the earth being towards sun's instantaneous location and not where it looks to be, at any given point in time. The motion of the earth around the sun and how the earth experiences the sun's gravity is graphically shown below.

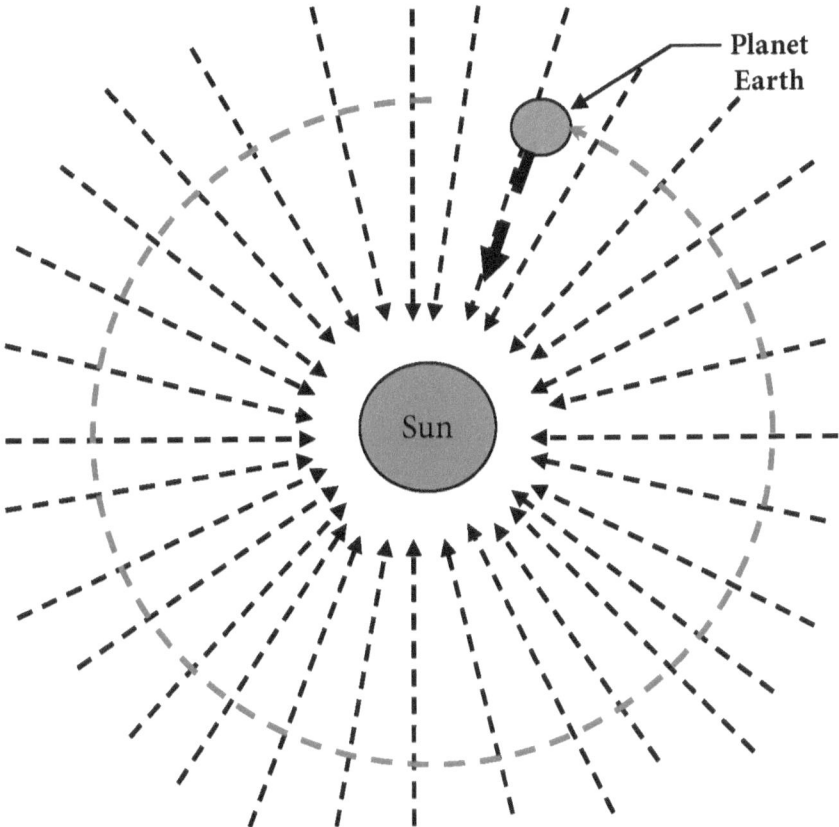

The flow of aether is always directly towards
the instantaneous location of the sun.

The flow of aether in space follows a straight path, towards a star or even a planet. Any object that enters this flow, it is affected instantly, and not after a few seconds or minutes. It instantly experiences the drag force that is induced by the flow. The exerted force is always directly towards the current location of the planet or star.

Note that, the flow of aether due to the existence of planets and stars is always there, regardless of the other physical objects being present or not.

In fact, the sun and the earth (and all other celestial bodies in this universe) don't know that any other celestial body of any size even exists. At any given instant, any celestial body literally independently generates an influx of aether towards itself, a flow that induces a drag force on objects that happen to be in its path. This drag force manifests itself as the force of gravity that is experienced by all objects in this universe.

10- The widening of the planetary orbits around the sun

As the aether's density is decreasing over time, due to its expansion as well as leakage into the accompanying universe, it is becoming less effective in dragging objects that happen to be in its path.

Therefore, stars are gradually, apparently exerting less of a gravitational force on their respective planets. This simply means that stars are gradually losing their grip on their planets and consequently,

"The planetary orbits are gradually becoming wider."

This side effect of decrease in aether's density has already been detected (in regards to the planet earth). According to the collected data, the orbital path of the planet earth is widening by about 7 meters (about 23 feet) per century.

Of course, a small portion is justifiably due to the sun losing mass as it is continuously transforming some matter into energy and also as it is literally throwing some matter into space as solar storms.

11- No direct interaction between aether and matter

Properties of aether are explained in great detail in a separate section of this book ("What is aether?"). To summarize, as far as it is related to this section,

Even though matter is one of various ways that aether manifests itself, aether does not interact with matter, other than dragging it along its path.

Matter and anti-matter can even be considered as chunks of ice that are literally dragged by the flow of water in an ocean.

12- The Michelson and Morley experiments

The Michelson and Morley experiments are explained in detail in a separate section of this book, "Michelson and Morley Experiment (Revised)". A brief summary is given below.

The Michelson and Morley experiments were all performed on a level setting, since they were expecting that the earth is literally floating in a medium that is stationary. They had not taken other affecting phenomena such as the force of gravity associated with the earth, the moon, the sun and the center of the galaxy into account. Therefore, their apparatus could not perform the task for which it was intended.

However, if such experiments are repeated in the future, while taking into account the important points that are clearly stated in this book, the apparatus would detect the motion of the earth relative to aether, as well as the effects associated with the gravitational forces of the earth, the moon, the sun and the center of the galaxy.

13- The initial sudden and yet temporary expansion of the universe during its infancy.

As it is presented in this section, the force of gravity is literally the drag force that is induced by the flow of aether towards drain holes, namely the matter and anti-matter particles.

Since there were no matter and/or anti-matter particles present when the initial sudden expansion started, even though aether was experiencing its highest pressures ever, it could not experience any such force as gravity, because **there was no aether flow**. Even during its sudden expansion process, it was not confronted by any kind of resistive force such as gravity. That is why the expansion process went on for awhile until the density of aether was reduced to a certain level that allowed the formation of matter and anti-matter particles.

Matter and anti-matter particles were literally holes that were formed in the fabric of space (just like tiny tubes) through which aether could escape this universe and enter the accompanying universe. That was when and how the force of gravity was actually born. In other words,

"The very formation of matter and anti-matter particles marked the Birth of the Force of Gravity."

The very formation of matter and anti-matter particles and their sudden abundance to the point of saturating the whole medium of aether caused two major side effects:

1- Due to their abundance, matter and anti-matter particles allowed unrestricted flow of aether from this universe into the accompanying universe. In a relatively short period of time, the aether pressure was reduced drastically, the same pressure that was the driving force for aether's expansion.

2- Due to their abundance, matter and anti-matter particles generated/ imposed such a strong gravitational force towards each other that literally acted like a braking system and slowed their expansion process in the whole universe. As most of the matter and anti-matter particles united and annihilated each other, leading to severe reduction in the force of gravity present, the leftover matter particles continued with their expansion process in this universe, but at a much slower pace.

"The formation of matter particles (in general) literally drastically slowed down the spreading of the contents of the universe, and the expansion of space."

In other words,

"The very function performed by the matter and anti-matter particles was literally like <u>an effective braking system</u> that drastically slowed the sudden expansion process of the universe, once it had reached a certain stage."

Conclusions

According to the new theory presented in this section, aether exists. Its flow towards each and every sub-atomic particle, all of which act as drain holes, induces a drag force on all other particles and hence objects of all sizes in its path. The drag force experienced is in fact the force of gravity which everyone is familiar with.

One has to be reminded that aether is not a stationary medium as it was assumed by the 19th century physicists such as Mr. Lorentz, Mr. Maxwell, Mr. Michelson and Mr. Morley. It is a dynamic medium. The variety of the flow patterns that concurrently exist in aether, due to the forces of gravity associated with a variety of objects and even celestial bodies, as well as the local magnetic and electric fields, can be visualized just like the variety of the flow patterns that are constantly experienced by the air molecules in the atmosphere or by the water molecules in the oceans.

In summary,

"The force of gravity experienced at any given location in space is due to the drag force induced by the net flow of aether."

One should also note that, as aether is escaping through matter and anti-matter particles (as well as black holes), the pressure and hence the density of aether in this universe is decreasing. This in turn means that aether is becoming less effective in inducing the drag force which is the force of gravity. Therefore, as time goes on, the force of gravity is gradually becoming weaker. Eventually, as the aether pressure in this universe and the aether pressure in the accompanying universe equalize, the force of gravity will literally fade away. In other words,

"Just as the very formation of matter and anti-matter particles caused the birth of the force of gravity, their continued existence will eventually cause it to vanish."

Gravity and Light

Gravity and Light

This section aims to demonstrate that gravity directly affects the path of light and other electromagnetic waves. This is not because there are any curvatures in the spatial dimensions but rather due to the flow of aether, the medium through which light travels in the form of phase vibrations.

In 1865, Mr. Maxwell proposed his electromagnetism theory. Using his theory he quite accurately calculated the speed of light in a vacuum. Mr. Maxwell had based his theory on the existence of some kind of medium through which light and other electromagnetic waves were propagating. This medium at that time was referred to as <u>Aether</u>. What aether was made of, or what kind of properties it had was not known. It was just necessary to be there as a medium for electromagnetic waves to travel in. The aether medium to electromagnetic waves was thought to be just as air is to sound waves.

Scientists of the time had thought of aether as a **stationary medium** which carried electromagnetic waves.

Mr. Michelson and Mr. Morley designed and performed the most promising experiments to detect the relative motion of earth and aether, as earth follows its path in space. Even though they conducted their experiments under a variety of conditions, no relative motion was ever detected.

Note that, Michelson and Morley's experiments were designed to detect relative motions between earth and aether on a level / horizontal plane.

They did not detect any horizontal flow due to earth's motion because the speed of aether flowing in the vertical direction is much faster than its speed due to earth's motion in space.

Over the last one hundred years, various international scientific groups have confirmed the following findings in their reports regarding light in this universe:

1- As light passes close enough to a galaxy (or even a star), its path bends towards the galaxy (or the star).

2- If light passes close enough to a black hole, it will be pulled into the black hole.

3- The farther the galaxies are from earth, the faster they are getting away from it. The rate at which a given galaxy is receding from earth is calculated using the red shift in the light spectrum received from that particular galaxy.

According to these findings:

1- Gravity can change the direction in which light is propagated.

2- Gravity can literally stop the motion of light.

3- Relative motions of the source and the observer only affect the frequency of the light received.

4- According to astrophysicists and physicists, the speed of light in a vacuum is constant and independent of the relative velocities between its source and its receiver.

Every single effect mentioned is a duplicate of the way sound is affected as it traverses through the air.

According to the theories that are presented in this book, aether is responsible for a variety of effects and forces in nature. The behaviors associated with the force for gravity, the magnetic field and electric fields, among other things, are literally due to a variety of motions in the aether medium.

Therefore, in order to isolate and detect the flow of aether that is actually inducing the force of gravity, one needs to conduct his experiments at certain locations and at certain times.

The experiment described below is designed to demonstrate how the force of gravity which is actually the drag force induced by the flow of aether, affects the path of a light beam.

The design details of the apparatus to be used for this experiment are shown below. The main components of this setup are:

1- A strong light source.
2- A lens to focus and linearize the light beam.
3- One flat mirror with single reflecting surface.

4- One convex (2-dimensional) mirror with single reflecting surface.
5- A flat screen that is calibrated 2-dimensionally.
6- Two main test stands/frames. One at the source of light and the other at the receiver end, 10 km or more away.

North

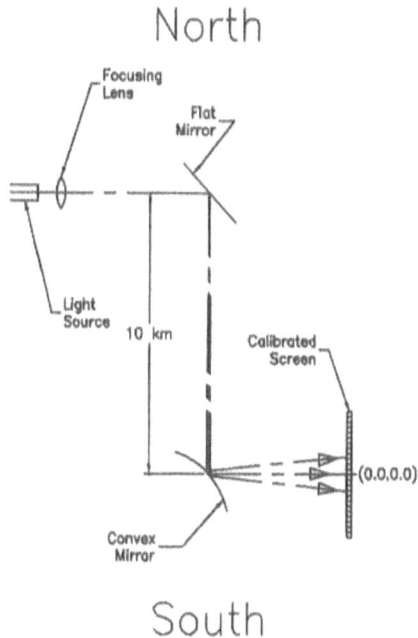

South

The North/South direction indicated in the figure is preferred, but it is not necessary.

In North/South, East/West or any other directions, the experiment will allow the collection of the very same data in regards to the gravity. The only difference will be in the shape of the resulting chart as the collected data will be plotted, both in regards to gravity and the earth's motion in space.

If the North/South direction is used, then the effect of gravity (as well as the earth's motion) will demonstrate a continuous and gradual shift between a minimum (at 12:00 noon), maximum (at 12:00 midnight) and again a minimum (at 12:00 noon).

However, if any of the other directions are picked, particularly the East/West direction, then the effects due to earth's motion will result in an oval shape chart, but with maximum and minimums still occurring at the very same times as the North/South setup.

To show the effect of aether's flow that induces the force of gravity, this experiment has to be performed under certain conditions and more importantly at a specific location and during a specific time period. The required criteria are:

1- The experimental apparatus must be setup so that the line connecting the two main test stands/frames is as close to horizontal as possible.

2- The experiment should be setup at relatively high altitude, due to less air pollution and less spreading/dispersion of the light beam used.

3- The light beam should be (preferably) travelling in the North-South direction.

4- Having prior knowledge of the ocean's tides being caused by the gravitational pulls of the moon and the sun will help to clarify the reasoning for the following more restrictive criteria. This experiment must be performed:

- Over a continuous period of 27 hours, between 10:00am and 1:00pm.
- When the moon is between the earth and the sun and nearest to the line connecting their respective centers.
- When the earth, the moon, the sun and the center of the Milky Way galaxy are in line, in the order given.
- At a suitable location on the equator, or as close as possible.

It is also important to mention that:

1- This experimental setup may be assembled at any time of the day or night.

2- The final adjustments and fine tuning of the light beam must be done when the test is to be performed.

3- The calibrated screen must be oriented so that one of its axes will demonstrate the deflection of the light beam in the vertical direction towards the earth. This can be achieved by vertically adjusting the light beam from up to down and rotating the calibrated screen to the desired orientation.

When the center of the galaxy, sun, moon, and earth (in that order) are all in line (leading to accumulative gravity effect), the experiment must be conducted at a location where it is in fact in line with the centers of the earth, moon, sun and center of the galaxy. This flow of aether will be vertical to the surface.

This experiment must start at about 10:00 am. The direction of the light beam from the source to the flat mirror, to the convex mirror and then to the calibrated screen should be adjusted so that the light spot shown on the calibrated screen hits the coordinate position (0.0, 0.0) as defined by the x and y axes, right at 12:00 noon. Then, as time goes on, the position indicated by the light beam on the calibrated screen must be recorded every hour, starting at 12:00 noon through 1:00pm, the next day. This length of time period will cover the two extremes of the strength of the gravity, at that particular location.

At 12:00 noon the apparatus will register the effect of the force of gravity exerted by the earth <u>minus</u> the forces exerted by the moon, the sun and the center of the galaxy put together. However, at 12:00 midnight, the apparatus will register the effect of the maximum force of gravity exerted by the earth <u>plus</u> the forces exerted by the moon, the sun and the center of the galaxy.

Note that, at 6:00pm and 6:00am the apparatus will also register the deviations caused in the direction of the light beam by the combined forces of gravity exerted by the moon, the sun and center of the galaxy, as they will be at 90 degrees to that of the earth.

Over the period of the experiment, the deflected light beam will hit the convex mirror at different points, forming a circle. However, due to the nature of the convex shape of the mirror, the angle at which the light beam will be reflected will be exaggerated.

The ideal timing and therefore positions of the earth, the moon, the sun and the center of the galaxy needed to perform this experiment is indicated in the figure below.

The timing and position of the experiment

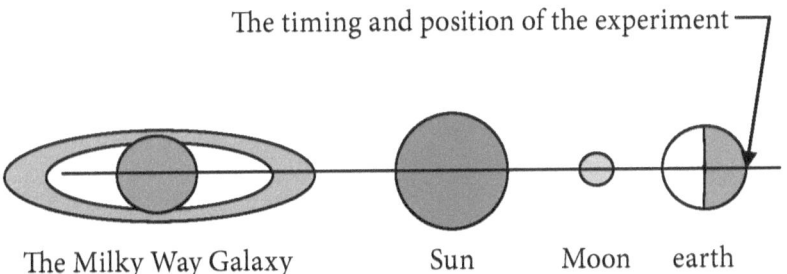

The Milky Way Galaxy   Sun   Moon   earth

Note that, Using the data collected during this test (the amount of deflections detected/recorded at different times), and also knowing the overall speed at which earth is travelling through space, one can readily calculate the speed of the aether that is causing the combined gravity forces associated with the earth, moon, sun and center of the galaxy.

What is Magnetic Field?

What is Magnetic Field?

Mr. Maxwell proposed his theory on electromagnetism, in 1865. Using his equations he was able to quite accurately calculate the speed of light in vacuum. Mr. Maxwell had based his theory on the existence of some kind of medium through which light and other electromagnetic waves were propagating. This medium at that time was referred to as Aether. What aether was made of, or what kind of properties it had was not known. It was just needed to be there as a medium for electromagnetic waves to travel in. To electromagnetic waves the aether medium was thought to be just as air is to sound waves.

On the contrary to what was thought back then, the aether medium is not a stationary medium. The varieties of motions that exist in the aether medium can be likened to the motions of the air molecules in the atmosphere or the motions of the water molecules in the oceans.

This section is specifically aimed at demonstrating that magnetic force field is in fact one such motion that is induced in aether. To do so it is providing the details of two experiments that will independently demonstrate how the magnetic field generated by a strong magnet affects the propagation of a light beam (electromagnetic waves) and hence reveal/confirm the essence of the magnetic field.

One experiment shows the cross flow effect and the other shows the direct inline flow effect. The cross flow effect is demonstrated using a very long horseshoe type magnet setup, while the inline flow effect is demonstrated by using a long coil type magnet.

In both cases, the experimental setups as well as their respective procedures are clearly explained.

First experiment: The effect of magnetic field on the direction of light (the crossflow effect)

This experiment is designed to demonstrate the effect of the magnetic field on the direction travelled by light or any other electromagnetic wave.

The basic idea in this experiment is to demonstrate, how the motions induced in aether by a magnetic field can drag the light waves downstream and hence cause a shift in the projection of the light observed on a calibrated screen.

The design details of the apparatus used for this experiment are shown below.

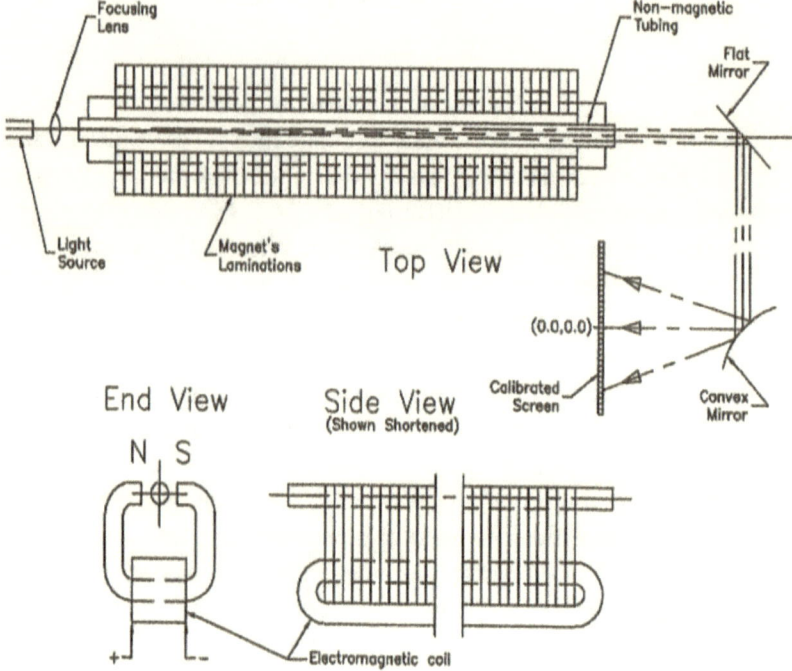

The main components of this setup are:

1- A strong light source.
2- A lens to focus and linearize the light beam.
3- A long electrically energized horse-shoe shaped magnet with the poles directly facing each other.
4- One long tube made of non-magnetic material.
5- Electrical power supply (Direct Current, DC) with a wide range of output voltages.
6- One flat mirror with single reflecting surface.
7- One convex mirror with single reflecting surface.
8- A flat screen, calibrated 2-dimensionally, <u>located as far as possible from the convex mirror.</u>

It is important to mention that:

1- This experimental setup may be assembled at any time of day or night.

2- The final adjustments and fine tuning needed for the light beam must be done when the test is to be performed.

Note that, depending on the sensitivity of the equipment, if the same adjustments are to be used at different times and even different locations, the effects of the other natural phenomena such as the gravity and the motion of the earth around the sun and the solar system in the galaxy must be taken into account.

3- The calibrated screen must be oriented so that one of its axes will demonstrate the deflection of the light beam parallel with the magnetic field lines of the magnet.
   This can be achieved by adjusting the light beam from side to side (between North and South poles), while the magnet is not activated, and rotating the calibrated screen to the desired orientation.

4- This experimental apparatus must be fine tuned without the magnetic field being activated. The light beam must be properly directed through the tube, which is located between the poles of the magnet. It must be adjusted so that it is focused on the (0.0, 0.0) point of the calibrated screen.

5- This experiment is to be repeated with different magnetic field strengths. This can be achieved by using different voltages supplied to the magnet's coil.

This experiment should be started without activating the magnet. In this case, the light beam will pass through the tube without being affected. The light spot shown on the calibrated screen will be centered right at the (0.0, 0.0) position.

Next, using the minimal voltage setting available, the electric magnet must be turned on. The induced magnetic field will cause the light beam to deviate from its straight path. The amount of this deviation will directly depend on the strength of the magnetic field generated. This effect can be easily demonstrated by energizing the electric magnet with different voltages.

The deflected light beam will hit the convex mirror at different points. Due to the nature of the convex shape of the mirror, the angle at which the light beam will be reflected will be exaggerated. Also, due to the long distance between the convex mirror and the calibrated screen, the light beam will show its presence at different spots on the calibrated screen, as the magnet is energized by different voltages.

This experiment can also be repeated with the poles of the magnet reversed. This can be achieved by simply reversing the wires supplying the electrical power to the magnet's coil. The change in the direction of the magnetic field will automatically reverse the light show observed on the calibrated screen.

Second Experiment: The effect of magnetic field on the speed of light (the direct inline effect)

This experiment is designed to demonstrate the effect of magnetic field on the speed of light or any other electromagnetic wave.

The design details of the apparatus used for this experiment are shown below.

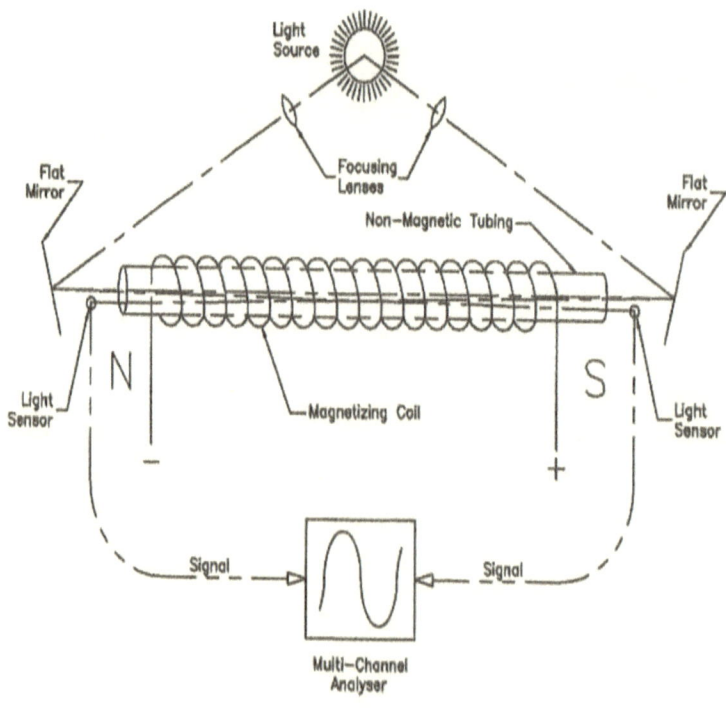

The main components of this setup are:

1- A strong monochromatic light source.

2- Two identical flat mirrors with single reflecting surfaces.

3- One long cylinder made of non-magnetic material. This cylinder is wound with many turns of an insulated wire. When activated with an electric current, this coil of wire will induce a magnetic field along the length of the tube.

4- Electrical power supply capable of delivering a wide range of Direct Current (DC) voltages.

5- Two identical super sensitive light sensors.

6- One multi-channel analyzer-scope capable of concurrently showing the exact sine wave shapes as well as the frequencies of the signals received from the two light sensors.

It is of great importance to mention that:

1- This experimental setup may be assembled at any time of day or night.

2- The final adjustments and fine tuning needed for the light beam must be done when the test is to be performed.

Note that, depending on the sensitivity of the equipment, if the same adjustments are to be used at different times and even different locations, the effects of the other natural phenomena such as the gravity and the motion of the earth around the sun and the solar system in the galaxy must be taken into account.

3- While the coil is not activated, the apparatus must be calibrated for several effects:

• Since the two light sensors are receiving light from the very same source, the frequency of the light detected will therefore be identical.

- Since the two light sensors are identical, and the intensity of the light beam reaching both are also the same, the amplitude of their output signals should be very close to each other. Using the electronics one must adjust and match their amplitudes.

- The timing of the two signals shown concurrently on the scope must be synchronized. This required fine tuning may be achieved either electronically or by adjusting the lengths of the two wires delivering the signals from the light sensors to the multichannel analyzer.

4- This experiment is to be repeated with different magnetic field strengths which can be achieved by using different voltages to energize the magnet.

This experiment should be started without activating the magnet. In this case, the two sine waves representing the signals received from the two light sensors will be superimposed on each other.

Next, using the minimal voltage setting available, the electric magnet must be turned on. The induced magnetic field will cause a difference in the speed at which the two light beams will propagate in opposite directions.

The presence of the magnetic field will speed up the light beam travelling in one direction, while it will slow down the other beam travelling in the opposite direction. Therefore, the two sine waves corresponding to the signals received from the two light sensors, shown concurrently on the scope, will become out of phase. They will be shifted by equal amounts in the opposite directions.

Note that, the shift in the sine waves will be only in the timing of their signals and not their frequencies. There will be no red shift or blue shift detections in these experiments. This is due to the fact that, the emitter and the receivers are stationary relative to each other.

The amount of this induced difference between the speed of light beams flowing with and against the magnetic field lines, and the apparent shift in the position of the two sine waves on the scope, will directly depend on the strength of the magnetic field generated by the magnetic coil, as it is energized by different voltages.

Note that, since the multichannel analyzer can also provide the frequency of the light signal used, one can create a chart demonstrating the amount of the shift in the sine waves, relative to each other, as a function of the strength of the magnetic field generated as different voltages are used.

The amount of shift generated in the two sign waves, as compared to the width of a full wave will allow the calculation of the speed difference induced in the two light beams travelling in the opposite directions through the magnetic field of certain strength.

The amount of shift recorded will provide the information needed to determine the strength of the magnetic field necessary to slow the light beam by any desired amount.

By using a magnetic coil composed of a greater number of turns and energizing it with the highest voltages necessary, one can even try to literally stop the forward movement of the light beam. The light beam (wave) have to fight its way upstream against the force of the induced magnetic field which is in fact a form of motion that is induced in aether.

Conclusions

It has been explained how one may experimentally demonstrate that the magnetic force field is in fact a form of motion induced in aether. This type of motion in aether actually forms a complete loop, a loop that can be quite small (such as the magnetic field lines around a regular magnet) or quite large in scale such as the field lines around / on / through a planet or even a star.

Note that, the earth has its own magnetic field, just as the sun and other celestial bodies do.

The effects of magnetic fields on a beam of light are shown in both cases of cross flow as well as direct inline flow.

Michelson and Morley Experiment (Revised)

Michelson and Morley Experiment
(Revised)

This section aims to demonstrate that the Michelson and Morley experiments would have indicated the existence of aether if they were performed differently and other affecting phenomena were also taken into account.

In 1865, Mr. Maxwell proposed his theory on electromagnetism. Using his theory he quite accurately calculated the speed of light in a vacuum. Mr. Maxwell had based his theory on the existence of some kind of medium through which light and other electromagnetic waves were travelling. This medium at that time was referred to as <u>Aether</u>. What aether was made of, or what kind of properties it had was not known. It was just needed to be there as a medium for electromagnetic waves to travel in. The aether medium to electromagnetic waves was thought to be just as air is to sound waves.

Over time the scientists became curious about aether and tried to understand its behavior. They thought if aether was stationary in space and all waves and even planets were passing through it, then one should be able to detect its drift velocity, on the surface of the earth. Knowing that the earth was orbiting the sun at about 107,000 kilometers per hour, various attempts were made to detect this drift velocity.

The most promising of all the experiments were designed and performed by Mr. Michelson and Mr. Morley during 1887. Again, these two gentlemen had assumed that aether was stationary in space and earth was moving through it. They had designed their apparatus with sufficient accuracy to detect such a minute velocity variation caused by the earth's motion around the sun.

Their experiment was based on the comparison of wave patterns created by the same monochromatic light passing through two paths, one in the direction of the earth's motion around the sun and the other at 90 degrees to the said direction. By reflecting back and superimposing these two light branches, Mr. Michelson and Mr. Morley were hoping to see the formation of some sort of interference pattern between the two.

They tried their experiment on the surface of earth, at various locations. They also conducted their experiment up at high altitudes, with the help of a balloon. However, no interference pattern was formed in any of their attempts. Some scientists suggested that the planet earth, as it goes around the sun, can be carrying a thin layer of this aether medium with it. This would explain why no effect was detected.

In 1904, Mr. Fitzgerald and Mr. Lorentz, independently, provided reasoning for detecting no trace of any drift effect in all of the experiments. Mathematically, they proved that the length of the experimental apparatus's arm which was in the direction of the earth's motion would have been affected by the motion itself and therefore had resulted in annulling the intended effect of the aether drift.

Mr. Lorentz, in 1904, proposed a theory "Principles of Relativity" based on which the effect of earth's motion on the apparatus was clearly explained. This theory was the first theory to ever talk about the shortening of the length of the object in the direction of motion, as well as the slowing down of the passage of time, as one approaches the speed of light in a vacuum.

Note that, Mr. Lorentz and Mr. Fitzgerald had proposed their theories without rejecting the existence of aether which was commonly accepted by their contemporary scientific community.

Also, the experiments performed by Mr. Michelson and Mr. Morley never indicated that there was no aether in space, or that the speed of light was the maximum speed possible.

In 1905, Mr. Einstein used Mr. Lorentz' theory and added two extra assumptions to it:

1- Mr. Einstein rejected the existence of aether, all together.

2- Mr. Einstein claimed that the speed of light in a vacuum is the maximum possible speed and it is a constant which is independent of the observer's state of motion relative to the source.

However, since Mr. Zwicky published his report on gravitational effects of some kind of invisible mass surrounding galaxies in 1930s, some new terms have been added to astrophysicists and physicists' vocabulary, namely the dark matter and the dark energy. Dark matter apparently does not directly interact with known forms of matter.

Based on the observable gravitational effects of dark matter on the rotational patterns of galaxies and even solar systems, dark matter is literally everywhere in this universe and it is particularly more concentrated in the space surrounding galaxies. According to astrophysicists, the known universe consists of:

73%, dark energy,
23%, dark matter, and
4%, all known forms of energy and all known forms of visible matter.

Over the last one hundred years, various international scientific groups have confirmed the following findings in their reports:

1- As light passes close enough to a galaxy (or even a star), its path bends towards the galaxy (or the star).

2- If light passes close enough to a black hole, it gets pulled into the black hole.

3- The farther the galaxies are from earth, the faster they are getting away from it. The rate at which any given galaxy is receding from earth is calculated using the red shift in the light spectrum received from that particular galaxy.

According to these findings:

1- Gravity can change the direction in which light is propagated.

2- Gravity can literally stop the motion of light.

3- Relative motion of the source and the observer only affects the frequency of the light received.

4- According to astrophysicists and physicists, the speed of light in a vacuum is constant and independent of the relative velocities of its source and its receiver.

All of these facts point towards the conclusion that light is a wave which travels in a medium, just as sound waves travel in air. Every single effect mentioned above is a duplicate of the way sound is affected as it traverses through the air.

According to the theories that are presented here in this book, aether is responsible for a variety of effects and forces in nature. The force of gravity, the magnetic field and the electric field, among other things, are literally due to a variety of motions induced in aether. Therefore, as one tries to detect aether's motion due to one phenomenon he must also take other affecting factors into account.

The apparatus used by Mr. Michelson and Mr. Morley was set up on a level/horizontal plain (floating on a bath of Mercury). Also, their experiments were performed at different locations, literally without paying any attention to the other phenomena influencing the local flow of aether and hence directly affecting the task performed by their apparatus. In those conditions, no matter which two orthogonal axes (directions) were picked, there would not have been any observable effects, since the direction of the flow of aether which is close to vertical would be the same for both of the apparatus' arms.

The timing and the location of the experiment are also very important, due to the overall motions of the earth in space (in the galaxy), as well as the motions that are induced in aether due to the gravities of the center of the galaxy, the sun, the moon and the earth.

When the center of the galaxy, sun, moon, and earth (in that order) are all in line, and their gravitational effects cause **aether motions** that are complementary to each other, the experiment must be conducted at midnight, on a night when:

— The earth's rotation at that particular location,
— Earth's motion around the sun, and
— The motion of the sun around the center of the Milky Way galaxy

are all in the very same direction, (leading to **motions performed in aether** that complement each other's effects).

The two overall motions of earth relative to aether (one due to gravity and the other due to the physical motion of earth in space) will be at almost 90 degrees to each other. The one due to gravity would be perpendicular to the surface, while the one due to various motions of earth in space would be tangent to the surface.

Therefore, one arm of the Michelson & Morley experimental setup should be near vertical, in overall (resultant) direction of the two motions of aether relative to earth. The other arm needs to be horizontal (level to the surface) and yet perpendicular to both of those two types of relative motions. In other words, it should be in the North/South direction, or perpendicular to the plain of the galaxy.

The output signals from both arms should be connected to a multichannel analyzer capable of showing both sine waves simultaneously.

In these conditions, as well as with proper timing, being midnight, the experiment which would have one arm raised up towards vertical (but tilted towards the direction of earth's rotation around its own axis) and one arm horizontal (perpendicular to the plane of the galaxy), would clearly demonstrate the desired effect.

The drawings below show the general setup, as in timing, as well as the directions of the apparatus' arms, as it is proposed here.

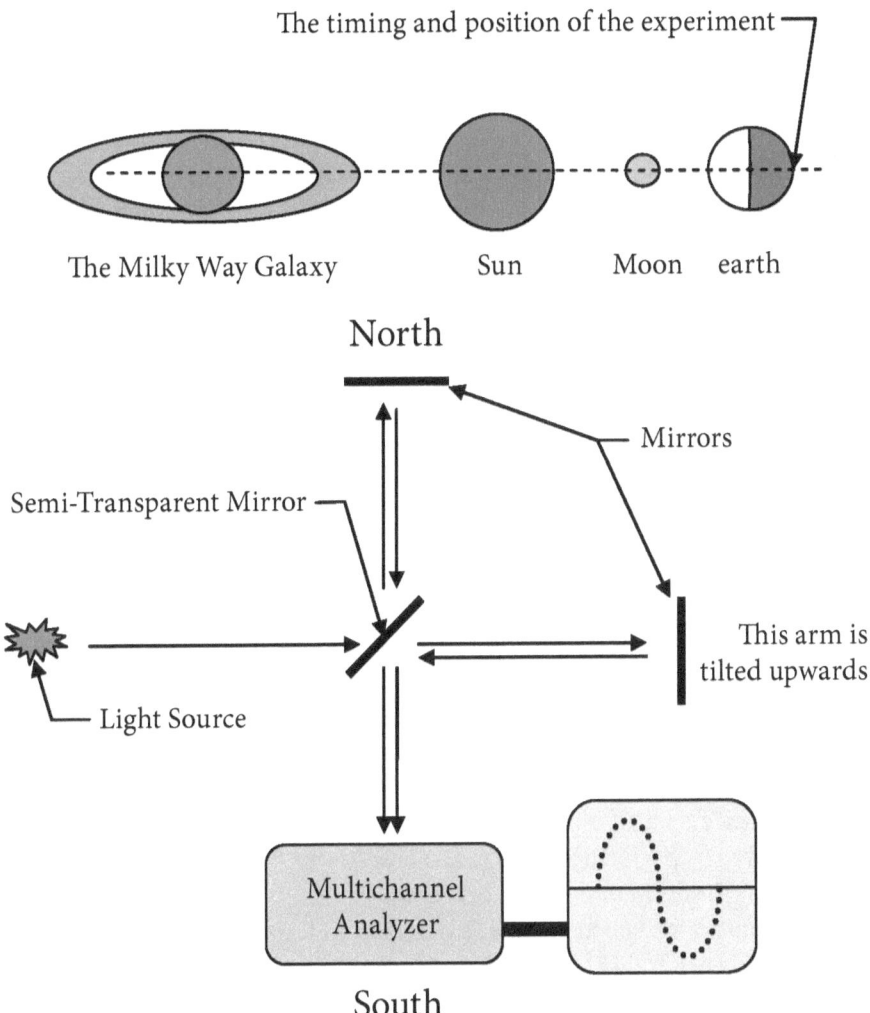

The timing and position of the experiment

The Milky Way Galaxy   Sun   Moon   earth

North

Mirrors

Semi-Transparent Mirror

This arm is tilted upwards

Light Source

Multichannel Analyzer

South

The accurate direction (vertical tilt angle) pointing towards the resultant of the two motions of earth relative to aether can be determined by adjusting the tilt until the maximum effect is observed.

Note that, using this tilt angle, one can readily calculate the speed at which aether is flowing due to the combined gravity forces present, as experienced at the time of experiment.

Also, the North / South direction indicated in the figure should be so that it takes into account:

— The tilt angle of the earth's axis relative to its orbital plain around the sun, and
— The tilt angle of the earth's orbital plain around the sun relative to the sun's orbital plain in the galaxy.

The signal that is traveling back and forth in the direction that corresponds to the highest speed relative to aether will result in **not a frequency change** but rather **a change in the length of the time of travel,** leading to a delay in the time of arrival of the signal.

The signal shown by the scope will indicate a shift in the location (timing) of the sine wave on the screen, but no changes in the frequency will be detected.

Note that, as one attempts to perform such experiments with the help of a multichannel analyzer and scope, he will no longer need to have two orthogonal arms and sophisticated split mirrors and precise timing of two beams of light and so on. **One arm of the apparatus would be all that is needed.**

By setting the single arm apparatus on a double axis pivot-mount (just like a gyroscope), one can point the device towards any desired direction and instantly see the amount of shift in the location of the sine wave generated.

One can also calculate the overall aether's speed relative to earth, theoretically. To perform such a task he needs to know the relative speeds due to earth's motion and due to gravity.

The rotational speed of the earth around its axis, and its various translational speeds due to orbiting around the sun and sun orbiting around the center of the Milky Way galaxy are known. However, the speed of the aether flow induced by the gravitational forces at that particular location and at that particular time has to be determined. This task can be performed in two ways:

1- By using the results from the section on "Gravity and Light".

2- By considering a hypothetical particle at infinity. As it literally starts to free fall directly towards a planet such as earth it gains speed due to the encouragements supplied by the aether flow towards earth, then the overall speed gained by the time the particle hits the surface should be a good indication of the local aether speed towards earth.

Black Holes And Their Properties

Black Holes and Their Properties

Black holes are defined as those regions in this universe that possess so strong a gravitational force that not even light waves can escape their grips, once they get close enough. Black holes are formed due to natural processes, as a result of massive stars imploding and all of their internal matter being compressed into a very small volume, while the outer layers are literally thrown into the surrounding space.

Various theories have proposed the existence of black holes, and have speculated on their characteristics and overall behaviors. However, there are many questions that need to be answered. The following are a few of such questions:

1- What really is a black hole?
2- How are black holes formed?
3- What is the minimum requirement for a celestial body to be called a black hole?
4- How is the force of gravity exchanged between a given black hole and matter particles or objects that may be nearby?
5- What does the event horizon of a black hole actually represent?
6- Can light stay in a stable orbit around a black hole?
7- How is time experienced by an object as it approaches and gets captured by a black hole?
8- Does time exist inside the event horizons of black holes?
9- What becomes of the particles or objects that are literally dragged into a black hole?
10- What is the final speed of any given matter particle or object as it falls into a black hole?
11- Can any information be extracted from the inside of a black hole's event horizon?
12- Are there different types of black holes?
13- Which black hole properties change as they grow by devouring more matter?
14- Can black holes split?
15- Can black holes be plugged?

16- Can black holes be kept in one place or be moved to another location?

17- Do black holes spin?

18- Do normal laws of physics apply inside the event horizons of black holes?

19- Do black holes gain weight just by sitting around and not devouring any matter particles?

20- Can black holes be neutralized?

21- How can the gravitational force of a given black hole be presented graphically?

22- Is there any artificial way of making a stable black hole?

23- What is the eventual fate/destiny of all black holes?

In order to understand what black holes are and what properties they might have, as well as be able to provide reasonable answers to the above questions, one needs to start with a sound theory on gravity.

What is Gravity?

The force of gravity is one of the main forces in nature. It is also the most common force that man has to deal with in his daily life. That is why, particularly over the last 400 years, man has developed sound theories regarding the laws that govern its behaviors.

Mr. Galileo proved that all objects regardless of their density fall towards earth at exactly the same rate. Mr. Keppler formulated the laws governing the motions of the planets around the sun. Mr. Newton, with the help of an apple, provided the basic laws of gravity.

Using Newton's theories and formulas, man has gained a remarkable amount of knowledge about the structure of the physical universe. Of course, especially over the last century, there have also been such theories as the general theory of relativity proposed by Mr. Einstein in 1917 that have tried to relate the force of gravity to the imaginary curvature of the 3 dimensions of the physical space.

One of the basic mysteries about the force of gravity is the speed at which it acts. All gravitational computations regarding the navigation of any inter-planetary probe is done using the instantaneous location of the planets and not where they look to be from the probe's position at any given point in time. Also, according to direct physical observations and data collected, at any given instant, the planet earth is pulled towards the true instantaneous location of the sun and not where it looks to be at that particular instant.

In other words, even though it takes about 8.3 minutes for sun's light to reach the earth, apparently it takes no time at all for the force of gravity to travel the very same distance. In short,

There are no gravitational theories that can explain the instantaneous effects of sun's gravity on earth and other planets.

The nature of gravity is explained in great details in a separate section of this book. A brief description of this new theory of gravity is presented below.

This new theory of gravity is based on two assumptions:

1- The dark matter which has been confirmed to exist everywhere in this universe, since 1933, is in fact aether, and

2- Any and all sub-atomic particles in this universe are in fact tiny drain holes (not black holes, but drain holes) through which aether, which is at high pressure inside this universe, flows into an accompanying universe that also has 3 dimensions and it is accessible only under certain conditions.

Note that, the accompanying universe is not a duplicate of this universe in any shape or form. It has its own duties to perform. Its existence enables this universe to function properly.

As a first step, for simplicity, consider one sub-atomic particle that is freely floating in space. The incoming flow of aether to this sub-atomic particle will be from all spatial directions. The flow will extend literally to the end of space, and it will always be directed towards the location of the subatomic particle. The flow of aether, towards this particle, is shown below (in a plane, due to simplicity).

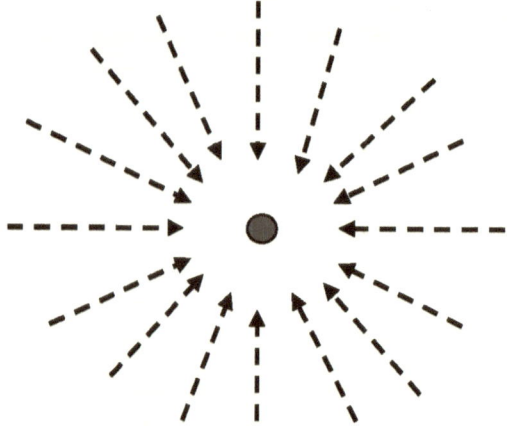

The flow of aether is directly towards the particle

As it is shown in the drawing below, any other particle would experience the effect of this induced aether flow as a drag force towards the sub-atomic particle.

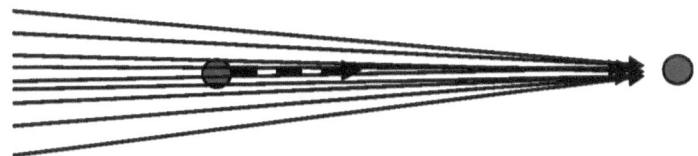

The drag force will be towards the true location of the sub-atomic particle.

**This drag force is the force of gravity that is experienced
by anything and everything in this universe.**

The flow of aether in space follows a straight path towards a star or even a planet. Any object that enters this flow instantly, and not after a few seconds or minutes, experiences the drag force that is induced by the flow. The exerted force will always be directly towards the current location of the planet or star. The flow of aether due to the existence of planets and stars has always been there and will always be there, regardless of the other physical objects being present or not.

In fact, the sun and the earth (and all of the other celestial bodies in this universe) don't know that any other celestial body of any size even exists. At any given instant, any celestial body literally independently causes a flow in the aether medium towards itself, a flow that induces a drag force on objects that happen to be in its path. This drag force manifests itself as the force of gravity that is experienced by all objects in this universe.

One has to be reminded that aether is not a stationary medium as it was assumed by the 19th century physicists such as Mr. Lorentz and Mr. Maxwell. It is quite a dynamic medium. The variety of flow patterns that concurrently exist in the medium of aether, due to the forces of gravity associated with a variety of objects and even celestial bodies, as well as the local magnetic and electric fields, are just like the variety of the flow patterns that are experienced by the molecules of air in the atmosphere or by the molecules of water in the oceans. In short,

**"The force of gravity experienced at any given position in space
is due to the drag force induced by the net flow of aether."**

Using this new theory of gravity, one can provide the following answers to the above questions.

1- What really is a black hole?

A black hole is basically an aggregate of immense number of matter particles, or aether drain holes to be more accurate, that are literally joined together. The aggregate is large enough to have a combined influx capability for aether that encourages the speed of aether approaching from the surrounding volume of space to reach the speed of light in aether.

This simply means that, any light (electromagnetic wave) that may reach such an aggregate can no longer escape and will be absorbed, because it will not be able to travel against the incoming flow of aether. Such a gathering of matter particles, due to being totally invisible, can be referred to as a black hole.

Note that, since black holes are aggregates of matter particles (drain holes for aether) that are literally joined together, they can be viewed as giant holes in this universe, that allow the aether to escape from this universe. In other words,

> **"Black holes are not a single point in space,**
> **they are meters or even kilometers across."**

2- How are black holes formed?

Black holes can form in at least two different ways, both of which are due to implosion of massive stars. They only differ in the intermediate steps, due to the size of the star.

As a massive star goes nova and literally throws part of its mass outwards into its surrounding space, it causes part of its interior to implode.

1- If the imploded portion is massive enough, it will allow aether to flow so freely that the incoming speed of aether will be in excess of the speed of light (speed of the phase vibrations in aether). This entity can be called a black hole, since light will not be able to travel against the aether flow, and will get dragged in.

   Note that, as a black hole is formed in one step, from a sufficiently massive star that has gone nova, the event horizon (explained below, question #5) will already be separated from the surface of the black hole.

   However, the initial distance between the event horizon and the surface will directly depend on the size of the imploded portion of the star. And, over time this distance can only get larger as the black hole will literally feed on whatever gets dragged in by the influx of aether.

2- If the imploded portion is not massive enough, it will become a neutron star, a star that is basically a solid mass of matter joined together. Over time, it gains more mass by attracting/absorbing matter from its surroundings. By continually gaining mass, neutron stars can eventually become sufficiently massive that they develop strong enough a gravitational force (appetite for aether) that the influx of the aether will reach the speed of light (or the speed of phase vibrations in aether). At this point, they can be officially referred to as a black hole.

> **Note that**, as a neutron star becomes a black hole, at first, its event horizon is literally right at its surface level. However, over time, as it gains more mass, its event horizon separates from its surface and spreads wider and wider into its surrounding space.

3- What is the minimum requirement for a celestial body to be called a black hole?

The minimum requirement for an aggregate of matter particles to be called a black hole is that their combined drainage effect should cause the speed of the aether flowing inside the aggregate to reach the speed of phase vibrations in the local aether medium.

In other words, for smallest black holes, the event horizon (explained below, question #5) is literally at the surface of the matter particle aggregate that forms the black hole.

However, as the black hole grows in strength, by devouring more matter particles, the event horizon separates from the surface and keeps on expanding into the surrounding space.

4- How is the force of gravity exchanged between a given black hole and matter particles or objects that may be nearby?

As it is briefly explained before (and in great details in a separate section of this book) the force of gravity is actually the drag force generated due to the motion of aether towards matter particles (which act as drain holes for aether).

Therefore, the force of gravity sensed due to the presence of a black hole or any other celestial body is not due to an exchange of force (or exchange of some kind of exotic particle) between the black hole and any matter particle (or object) that may be nearby. It is simply due to the drag force generated by the flow of aether towards the black hole or the celestial body which is made of a great many matter particles (aether drain holes) that are joined together.

5- What does the event horizon of a black hole actually represent?

Aether is constantly flowing directly towards black holes from all spatial directions. As the aether flow speeds up in its approach towards a given black hole it eventually reaches the speed of its phase vibrations which is the speed at which light and other electromagnetic waves travel through aether. This simply means that light waves will no longer be able to go backwards or away from the black hole and will get absorbed.

The distance (radius) from the center of a black hole at which the influx of aether reaches the speed of its phase vibrations is referred to as the **Event Horizon**. The event horizon can also be referred to as **the point (radius) of no return** for light or any other type of electromagnetic wave, as it approaches a black hole, since once it reaches the event horizon it can no longer go back.

The drawing below shows a black hole and its corresponding event horizon.

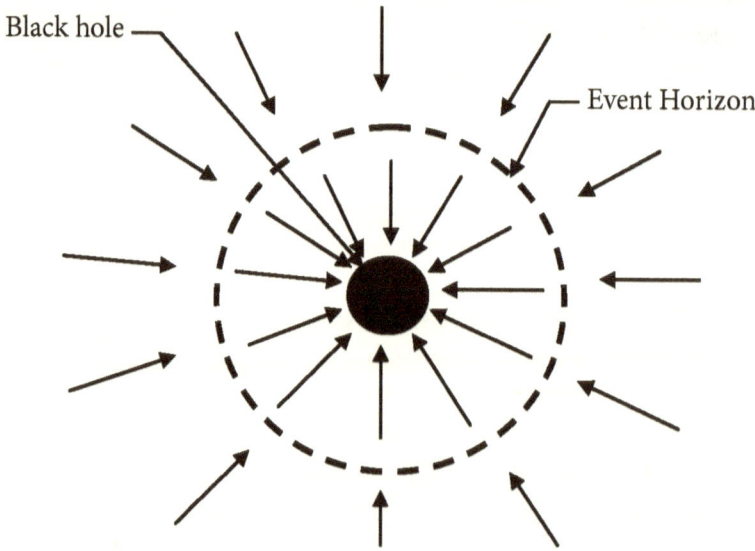

Arrows indicate aether flow from all directions.

<u>Note that,</u> light and other electromagnetic waves are phase vibrations in aether just as sound is a form of phase vibration in a medium such as air.

Therefore, as it is stated in the section on "What is Light?", speeds attainable by objects are not limited to the speed of light in a vacuum (vacuum being the aether medium that is void of any matter and/or anti-matter particles).

> **Therefore, even though light cannot get out of a black hole, due to its being literally carried by the aether flow, an object can fight its way out of a black hole, given it has the proper structural composition and some kind of aether propulsion system.**

6- Can light stay in a stable orbit around a black hole?

The answer to this question is, "Yes". However, for a light beam to stay in a stable orbit around a black hole it has to do so in an orbit that is outside of the black hole's event horizon, since if light reaches the event horizon of any black hole, it gets absorbed, automatically.

A light beam can stay in a stable orbit around a black hole, only if it is following a circular path around the black hole, at a distance so that its speed relative to the local aether that is flowing into the black hole is exactly equal to the speed of phase vibrations in that immediate aether medium. The drawing below demonstrates such an orbital path around a black hole.

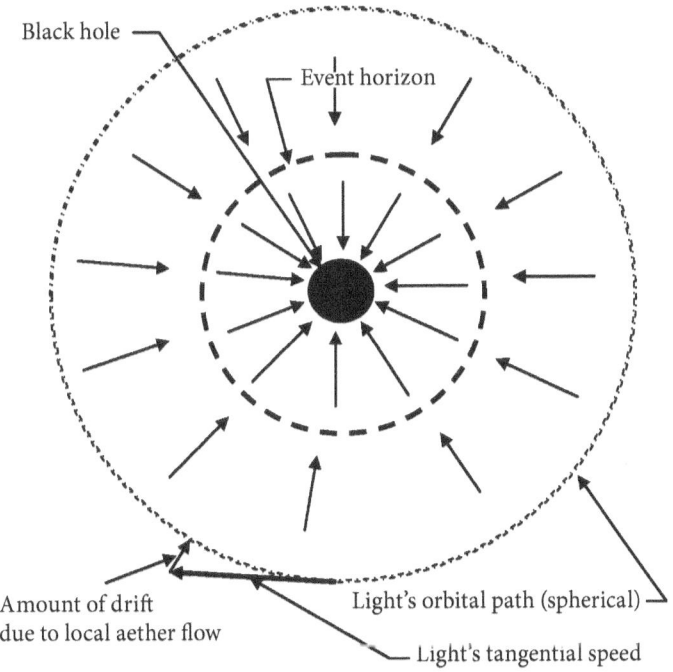

Black hole

Event horizon

Amount of drift
due to local aether flow

Light's orbital path (spherical)

Light's tangential speed

7- How is time experienced by an object as it approaches and gets captured by a black hole?

The essence of time is explained in detail in a separate section of this book. According to the information provided, any object that is moving at less than the speed of phase vibrations in the aether medium will always experience the passage of time. However, as its speed approaches that of phase vibrations in the local aether, it will experience time at a slower and slower pace. And, if its speed reaches the speed of phase vibrations in the local aether, the object will stop experiencing the passage of time, altogether.

Since black holes are literally massive drain holes for aether, aether is naturally constantly flowing directly towards black holes from all spatial directions. Therefore, the passage of time experienced by an object will depend on the manner in which it approaches a given black hole. There are two distinct methods of approach that can be followed by an object:

- **Tangential Approach**
  If the object barely notices the gravitational pull of a black hole as it is passing by, as shown below, it will not gain much of a speed, but its direction of motion becomes affected.

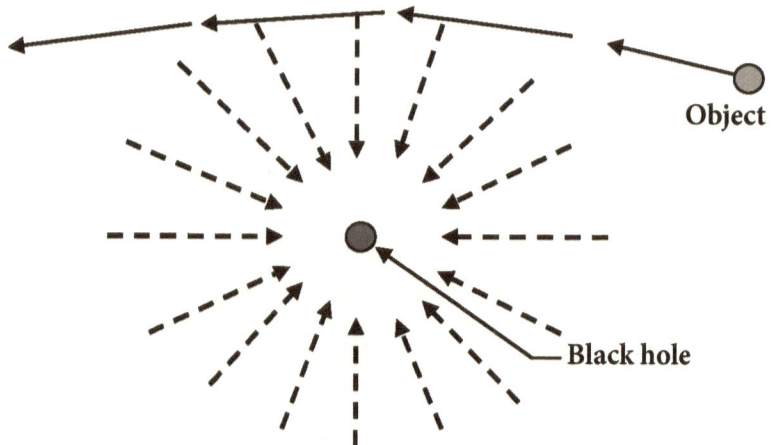

Arrows indicate aether flow from all directions

However, if the object passes close enough to the black hole, as it is shown below, it will get captured by the black hole's gravitational force. In this case, it will start a spiral motion around the black hole and will eventually get devoured by it. During its spiral motion, the object will continuously gain speed.

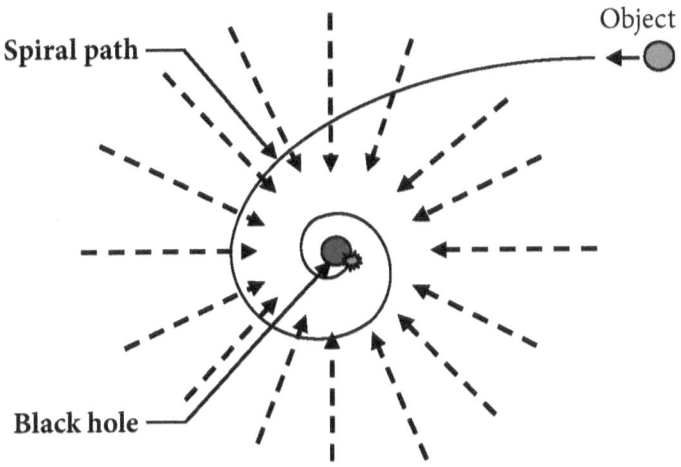

Arrows indicate aether flow from all directions

So long as the object's speed relative to the local aether that is flowing into the black hole stays below the speed of the phase vibrations in aether, it will experience the passage of time. But, the pace at which the passage of time is experienced will be increasingly slower due to gaining speed.

At some point, it will be moving at the speed of light which corresponds to the speed of the phase vibrations in the local aether. At that point, the object will stop experiencing the passage of time.

In fact, the object will reach this speed before it reaches the event horizon, since it will be moving literally almost perpendicular to the aether flow which is headed directly into the black hole. And, it is the overall resultant speed relative to the local aether that counts, as far as experiencing time is concerned.

In this case, the object that may happen to be some kind of a probe, it can send a light (electromagnetic wave type) signal towards the outside to warn the others or inform them of what it is observing / experiencing.

As the object follows its spiral path and crosses the event horizon and gets closer and closer to the surface of the black hole, its speed will be increased without limit. The only limitation is literally the length of the path that it has left to follow until it reaches the surface of the black hole.

In this type of approach to a black hole, the object will experience all of the special effects associated with relativistic speeds, because it will be moving relative to the aether and not with it.

Notes:

1- As the object speeds up beyond the speed of light (the speed of the phase vibrations in the local aether), the passage of time experienced will be zero.

2- By travelling faster than the speed of light, probes or matter particles (objects) will still be stuck to the present moment, continuously. In other words,

> ### "'The future' can only be experienced as it becomes 'the present'."

It should be mentioned that, any probe or object that continues to exist in its own present moment, as it is travelling faster than light, going directly away from its home planet will literally catch up with the light (electromagnetic waves) signals that were sent earlier, such as TV signals, regarding events that had occurred on its home planet.

These signals become detectable to the probe as it literally catches up with them. This simply implies that, the farther it goes (at super-light speeds) the older will be the signals that it will receive. In other words,

> "The information received will be backward in time, just like watching a video in reverse mode."

- **Direct Approach**
  If an object approaches a black hole directly head on, and allows itself to literally free fall into it, at all times, the object will be dragged freely by the aether flow (see drawing below).

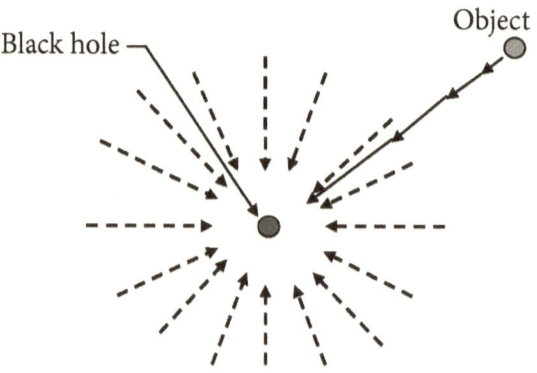

Black hole

Object

Arrows indicate aether flow from all directions

In this case, the object will have minimal speed relative to the local aether that is dragging it directly into the black hole. This is just like a piece of paper floating in a river stream. It too, will be almost at rest as compared to the water that is dragging it in any direction it flows.

In free fall mode the object will be moving at almost the same speed as the aether and in the very same direction. This means that it will reach the speed of the phase vibrations in aether, but without experiencing any type of time dilation effects, whatsoever.

Notes,

1- The occupants of a spaceship (or a scientific probe) that are headed directly into a black hole will keep on experiencing time at the very same pace as the individuals who are anxiously waiting to receive exciting news (telemetry) from them.

2- They will also be in their normal sizes, as well, since they will not shrink in the direction of motion. They will not experience any of the special effects associated with traveling at relativistic speeds.

3- The frequency of the transmission received by the bystanders will be at ever lower frequencies and will eventually cease, as the ship (or probe) reaches the speed of light and crosses the event horizon.

In this type of approach to a black hole, the object will not experience any of the special effects associated with relativistic speeds, because it will be moving with aether and not relative to it.

Note that, as the object speeds up due to being dragged by the aether flow, it will experience the speed of phase vibration as it crosses the event horizon.

Beyond event horizon, depending on the size of the black hole, the object will continue on speeding up until it reaches the surface of the black hole, but without experiencing any of the effects associated with relativistic speeds.

8- Does Time exist inside the event horizons of black holes?

As it is explained in a separate section of this book, "What is the Essence of Time?", time is a property of aether. The rate at which objects experience the passage of time is dependent on their motion relative to their local aether. The higher the speed relative to the local aether will result in experiencing time at a slower pace. As their speed relative to aether reaches the phase vibrations in their local aether, they will no longer experience the passage of time.

Therefore, in regards to whether time exists beyond the event horizon of a black hole or not, one has to state that it all depends on the relative motion of the object and the local aether, which is already flowing at higher speeds than the speed of light (speed of the phase vibrations in that medium).

As it is stated above, there are two distinct ways that objects can approach a given black hole. These two approach methods will result in totally different outcomes, as far as experiencing time is considered:

- **In the tangential approach method**, the speed of the object relative to its local aether medium reaches the speed of phase vibrations in aether, before it even crosses the event horizon.

  Therefore, objects that follow such an approach path will experience the passage of time at slower and slower pace as they gain more speed relative to aether. And, as their speeds relative to their local aether reach the speed of phase vibrations in the local aether medium, they will no longer experience the passage of time.

  They will experience this timeless state/effect starting right before crossing the event horizon and will continue to do so until they reach the surface of the black hole.

- **In the direct approach method**, which is in fact a free fall scenario, the speed of the object relative to its local aether medium is minimal, even as it crosses the event horizon of the black hole.

  Therefore, objects (probes) that follow such an approach path will not experience any of the effects associated with moving at relativistic speeds. These objects will continuously experience the passage of time at the very same rate as other objects (or beings) that may be far away and patiently waiting and hoping to receive telemetry data from their probes.

  Of course, the duration of such a period will be on the order of milliseconds, due to their high speeds and the distance that they have left to travel between the event horizon and the surface of the black hole.

9- What becomes of the matter particles or objects that are dragged into a black hole?

The step by step experiences of a given probe (matter particle) solely depends on the approach method used.

- **In the tangential approach method,** the probe will experience the extreme forces and distortions associated with moving at relativistic velocities relative to its local aether. In other words, it will get crushed, before it even reaches the event horizon of the black hole.

- **In the direct approach method,** the probe will stay intact, as it approaches the black hole, since it does not experience any of the special effects associated with moving at relativistic velocities.

  In other words, the probe will not experience any time dilation or the shrinkage of its longitudinal dimension. However, it will experience side pressures building up due to being funneled into a progressively narrower stream.

It must be noted that, one can only speculate about what happens to the matter after it reaches the actual surface of a black hole. A few of the possibilities are listed below:

1- The matter particles will blend in with the rest of the matter particles that are already there, and simply encourage the physical growth of the black hole and widen its event horizon.

   Note that, in this case, the overall mass of the black hole is increased by the amount of mass (matter particles) that it manages to devour.

2- All or part of the matter particles can be recycled by becoming of aether variety, and exit into another dimension, while the rest are absorbed locally and encourage the growth of the black hole.

   Note that, in this case, the overall mass of the black hole is increased only by the amount of mass (matter particles) that it has absorbed.

3- The matter particles (not objects) can literally pass right through the bottle neck of the black hole and exit at some other point in this universe, totally intact.

Note that, in this case, the overall mass of the black hole does not change at all. It only acts as an entrance or gateway into something like a wormhole that would connect to another part of this universe.

In this case, the black hole is literally acting like a port hole, although everything would be crushed as it passes through.

10- What is the final speed of any given matter particle or object as it falls into a black hole?

In the two approach scenarios mentioned earlier, namely the tangential and the direct approach methods, the speed of matter falling into a black hole will follow a slightly different pattern.

- **In case of tangential approach**, even though the speed of the object relative to the local aether reaches the speed of light before it crosses the event horizon of the black hole, but its actual speed will be less than speed of light, until it actually reaches the event horizon.

  As it continues, passed the event horizon, its actual speed will increase to even beyond the speed of light.

- **In case of direct approach**, the object is dragged directly into the black hole by the flow of aether which reaches the speed of the phase vibrations as it crosses the event horizon.

  However, because the speed of the matter is only due to direct and inline drag force exerted by aether, it will continuously be moving at a slightly slower speed than aether that is dragging it. Therefore, as it crosses the event horizon of the black hole it will be moving almost at the speed of phase vibrations in that medium. And, as it continues its journey towards the surface of the black hole, it will be gaining even more speed which means its **actual speed** (not relative to aether) will increase even past the speed of light in that medium.

Therefore, in both cases, the speed of any matter particle falling into a black hole will eventually be greater than the speed of light or the speed of the phase vibrations in aether.

Note that, the final speed of the matter particle falling into a given black hole will directly depend on the distance between the event horizon and the actual surface of that particular black hole. The greater the distance will result in more time to gain even more speed before reaching the surface of the black hole.

However, if the black hole is of the smallest possible size, which has its event horizon right at its surface, the matter or even the aether itself would not gain any extra speed as it crosses the event horizon.

11- Can any information be extracted from the inside of a black hole's event horizon?

As a probe may be sent into a black hole, using either of the tangential or the direct approach methods, one can expect different scenarios to be experienced.

- **In case of tangential approach,** the probe will take longer to reach the event horizon. But, its speed relative to the local aether will reach that of phase vibrations in aether long before that.

  In this case, the probe will be able to send electromagnetic type of signals towards the outside. But, before the probe even reaches the black hole's event horizon, it will get crushed due to losing longitudinal dimensional/structural integrity which means no more data will be sent by the probe.

- **In case of direct approach**, the probe directly approaches the event horizon and gains speed in the process, the telemetry will be at lower and lower frequencies. As the probe crosses the event horizon, even though the probe will stay intact as it gets closer to the black hole, as compared to the tangential method of approach, but no more signals of electromagnetic kind can be sent outward.

  Note that, if the probe is equipped with some kind of aether propulsion system, it will be able to break free from the black hole's gravity and head towards the outside. Otherwise, it will be dragged all the way in and be made one with the rest of the matter particles that are already there.

  It should be emphasized that, other types of propulsion systems will not be able to function in that environment. This is due to the fact that, the propulsion mechanism used must be able to literally manipulate the aether environment that it is in.

12- Are there different types of black holes?

Black holes can be categorized based on a variety of different criteria. But, in general, black holes may be categorized into the following six groups:

Stationary black holes: These black holes are mainly the ones that are formed in the central regions of galaxies. They are quite massive, due to having access to plenty of stars and ejected matter particles from them, literally from all directions.

Moving black holes: These black holes are formed at other locations all throughout the volume of the galaxies, as the individual stars explode/implode. The remnants of the stars that are turned into black holes will keep up with their respective parent stars' general motion in space.

Small/Young black holes: These black holes are either just recently formed or have been isolated from the rest of the matter contents of this universe, due to the positions of the stars that gave birth to them. The event horizons of these black holes are at or quite near their respective surfaces.

Massive/Old black holes: These black holes are mainly the ones that have formed in parts of the galaxies (including their centers) where the population density of stars are quite high. The event horizons of these black holes are totally separated from and are at large distances from their surfaces.

Spinning and Fixed black holes: All of the black holes start their lives with some initial spinning associated with them. This is due to the star's rotational momentum prior to going nova and forming the black hole, since as the mass of the internal portions of the star is squeezed into a very small volume, it causes the newly formed black hole to spin at quite a high rate.

Also, any and all types of black holes can encourage spinning actions in their surroundings. However, it is only the in-falling matter particles (stars and planets and so on) that are actually creating such a scene, as they are captured by the particularly massive black holes.

This is due to the fact that, only rarely a star or a planet happens to be literally aiming directly for their local black holes, to find out what all the fuss is about. They usually follow the tangential approach which automatically gives rise to their spiraling motions until they are finally devoured.

Note that, all of the black holes exhibit complex (3-D) spinning actions which are due to the arrival of the planets and stars from different directions, since different stars and planets do not arrive in a specific (common) plane. Each one contributes some rotation (angular momentum) in a certain direction of its choice, depending on its approach trajectory.

The spinning actions of black holes do not have any effect on planets or stars as they follow their crash-landing trajectories towards a black hole. These actions only affect them when they literally hit the surface.

If most of the stars and planets approaching a given black hole cause it to rotate in a certain direction, the faster the black hole will spin in that particular direction.

However, if enough stars and planets arrive at angles and directions that would cancel the black hole's original spin, the black hole will stop rotating altogether.

13- Which black hole properties change as they grow by devouring more matter?

As black holes grow in mass, due to the influx of aether literally dragging in more matter particles, not just their physical size becomes larger, but also and more importantly so, their event horizons cover wider regions of space.

The wider spread event horizon automatically means larger regions of space that can be affected by the black hole.

The wider spread event horizon also means that the influx of aether as well as the matter particles dragged in, will reach speeds that are much, much higher than the speed of light (the phase vibration in aether), before they reach the actual surface of the black hole.

14- Can black holes split?

Black holes are basically drain holes through which aether (not the matter particles, but aether) flows into another dimension. The influx of aether is the only reason why black holes even manifest their force of gravity.

Therefore, if somehow a given black hole gets plugged from its exit end, just like a plugged drain hole, even if for a short while, so that aether can no longer flow through, it will cease to act as a black hole. At this instant, even though the black hole may look like a chunk of matter (like a neutron star) but will have minimal to no gravitational force, due to no influx of aether. At this point, due to unknown forces, it may split literally into pieces. Each piece may become an active black hole later on, or if it is too small it would only act as a super condensed/compressed chunk of matter, kind of like a neutron star.

15- Can black holes be plugged?

From this end, this universe's dimensions, black holes are spherical entities. They accept aether from all spatial directions. Therefore, for a black hole to be plugged, from this end, it needs to be capped with a ball shaped material that can act as a barrier against aether flow.

Currently, there is no knowledge of the existence of such material, since aether can and does literally pass through planets and stars, let alone anything smaller.

Black holes are literally a passageway for the aether to go through to the other universe. Therefore, if somehow the flow is blocked within the black hole itself, it will cease to act as a black hole, just like a plugged drain hole.

16- Can black holes be kept in one place or be moved to another location?

Imagine a magnet (any shape or type) freely floating in the outer space. To manipulate its location, motion or even its orientation (directional effects), one needs to use something that will cause it to respond/react. By using another magnet or even a magnetically inclined piece of metal, at the proper distance, one can perform a variety of maneuvers in a controlled fashion.

The very same type of actions can be thought of as one considers black holes. Black holes like to devour matter particles. Therefore, to cause black holes to change their position and/or direction of motion, they need to be encouraged to go in the desired direction by literally being fed objects of different sizes from different directions and with different initial momentums.

Small black holes can be readily controllable in this fashion. For example:

- By sending a moon or a planet to a small black hole, with proper speed and at a predetermined angle, the black hole can be pulled or pushed towards the desired direction, as it is encouraged to literally feed on that planet. The drawing below demonstrates, in a simple form, how a planet (with initial speed/momentum) can be used to encourage a small black hole to change course and move in a specified direction.

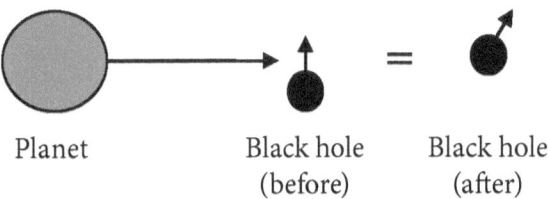

Planet   Black hole   Black hole
         (before)   (after)

Arrows indicate directions of motion

- A planet can be made to follow a slingshot trajectory to either stop the motion of a black hole, or at least slow it down, in a certain direction or promote its motion in a certain direction. As the planet makes its u-turn around the black hole, the black hole actually receives twice the planet's initial momentum in the planet's original direction of motion, since it has to stop its motion and give it a return momentum. The drawing below shows a simple presentation of such an undertaking.

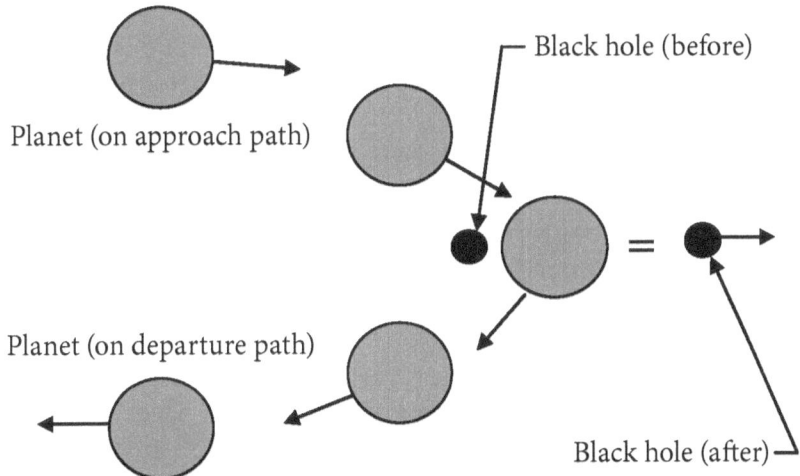

Arrows indicate directions of motion

The black hole will attract the planet, but will not be strong enough to pull it in. That is, if the minimum distance is properly chosen and the speed is kept above escape velocity at the nearest intended distance to the black hole. Therefore, the black hole will be literally pulled towards the desired direction.

However, in the case of giant black holes, it is generally wiser to simply leave them alone. In most cases, it is actually much easier to get out of their way, if one has enough of a prior notice/warning.

17- Do black holes spin?

All of the black holes start their lives with some initial spinning associated with them, due to the star's rotational momentum prior to its going nova and forming the black hole. When a star goes nova, the mass of its internal portions is squeezed into a very small volume, increasing its rotational speed and causing the newly formed black hole to spin at quite a high rate.

In general, all of the black holes exhibit complex (3-D) spinning actions which are due to the arrival of the planets and stars from different directions, since stars and planets don't arrive in a specific (common) plane. Therefore, each one contributes some rotational momentum in a certain direction, depending on its approach trajectory.

If most of the stars and planets approaching a given black hole cause it to rotate in a certain direction, the faster the black hole will spin in that particular direction.

However, if enough stars and planets arrive at angles and directions that would cancel the black hole's original spin, the black hole will stop rotating altogether.

18- Do normal laws of physics apply inside the event horizons of black holes?

So long as the motion of an object entering a black hole is in the form of voluntary free fall, the physical laws will still be valid, because, in this case the speeds attained relative to the local aether is minimal, even as the object crosses the event horizon of the black hole. But, **the laws of physics, as are currently known to man, will no longer hold once the object reaches the surface of the black hole.**

However, if the object is entering the black hole following a tangential trajectory, even before it reaches the event horizon, its speed relative to its local aether will reach that of phase vibrations in aether. This is due to the fact that, aether itself is already moving towards the center of the black hole at very high speeds, at near 90 degrees to the object's trajectory.

From the instant the object reaches the speed of phase vibrations in

aether, the laws of physics will no longer be valid, since time will become meaningless for that object. To apply normal laws of physics, one needs to experience time.

19- Do black holes gain weight just by sitting around and not devouring any matter particles?

The flow of aether into black holes is a continuous process, regardless of any matter particles or objects being present nearby to attract those black holes' attentions or not. Black holes and in fact every matter particle that exists in this universe, are continuously receiving aether from their respective surroundings.

Just as no matter particle such as a proton or a neutron is gaining any weight (mass) due to the continuous flow of aether into them, no black hole gets any heavier or more massive by devouring aether, either. Black holes gain weight (mass) only as more matter particles are literally dragged into them by the flow of aether.

Aether in its fluid state is kind of a diet food for matter particles and black holes. They all can and are literally swallowing aether, on a continuous basis, without gaining even one gram.

20- Can black holes be neutralized?

The only way to neutralize a black hole is by introducing it to another larger black hole. They would simply merge, without any hesitation. In fact, due to their enhanced mutual gravitational force, no news or details will leak to the outside world. In other words, the details of their joining process will be kept as a secret. The obvious effects will be the widening of the bigger black hole's event horizon, as shown below, and possibly a change in the direction of motion.

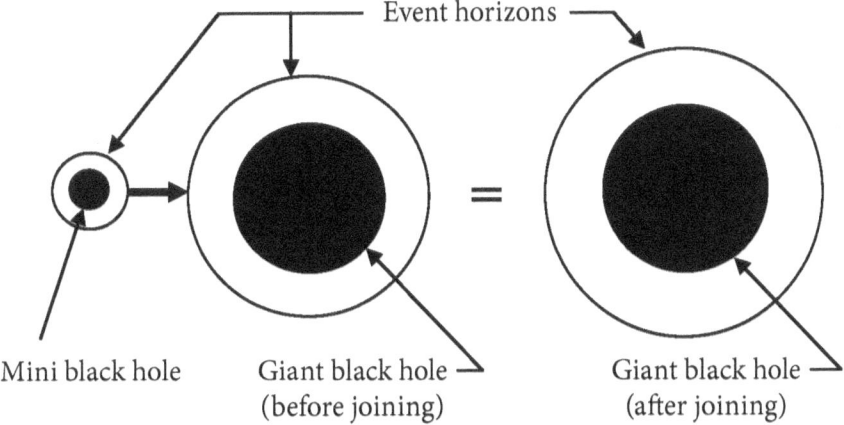

Event horizons

Mini black hole   Giant black hole
(before joining)

Giant black hole
(after joining)

21- How can the gravitational force of a given black hole be presented graphically?

The strength of the gravitational force of any black hole (in its surrounding space), as shown below, can be presented as a funnel type of a shape that has a smooth surface with gradually increasing slope towards its center.

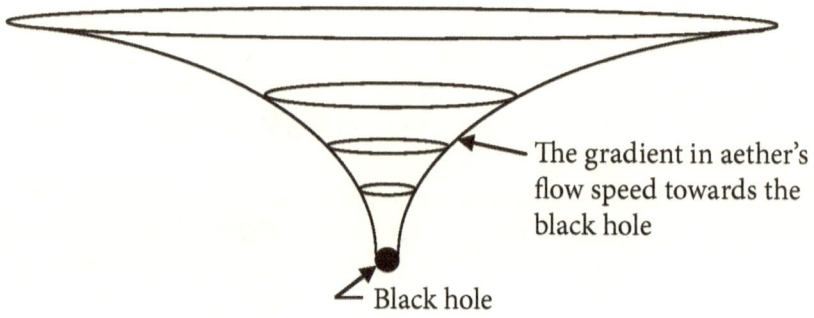

The gradient in aether's flow speed towards the black hole

Black hole

The flow of aether is from all directions

However, the gradual increase in the slope of the funnel shaped surface towards its center, shown in the figure, does not represent any kind of curvature in the spatial dimensions, but rather the gradient that exists in the ever increasing speed of the aether as it approaches the black hole's event horizon.

The very same is also true about the gravitational influences of other celestial bodies such as stars and galaxies. However, in their cases, the central portion does not become so pronounced in its slope.

22- Is there any artificial way of making a stable black hole?

The only way for a stable black hole to form is by the gathering of a great multitude of matter particles, as in a star, that goes nova. The other methods, magnetic/electric quenching, will only lead to temporary black hole effects which will only last for a moment. Therefore, the proper answer to this question is, 'NO'; there is no way to artificially manufacture a stable black hole. (Please refer to the section, "Predictions / Applications" for more details.)

Note that, underline unstable black holes or temporary black holes, which are in fact bypass routes to another location in this universe, can be manufactured artificially. They can be created by quenching/

imploding either electric field, magnetic field or a combination of the two, that are already strong, homogenous, stable and possess a certain frequency.

Even though these types of port holes allow the aether to flow through to the accompanying universe, just as black holes do, but their effectiveness is quite temporary. In order for them to continue with their intended functions, to transfer objects to the other dimensions and back (at some other desired location), they need to be energized, constantly.

These bypass routes can also be referred to as wormholes, portholes or shortcut routes. As an object enters one, it will exit through another which may be quite a distance away, even in a different galaxy.

23- What is the eventual fate/destiny of all black holes?

As the universe expands and the aether pressure in this universe as compared to the aether pressure in the accompanying universe approaches some **critical aether pressure difference**, all of the black holes, starting with the smaller ones, will cease to act as black holes. Because, once this critical aether pressure difference has been reached, the induced influx of aether reaching the surfaces of even the largest possible black holes will be flowing at speeds that are barely equal to the speed of the phase vibrations in aether. This critical aether pressure difference can be referred to as,

"Esmailzadeh Critical Aether Pressure Difference"

However, as the aether pressure continues to drop, and the speed of aether flow decreases, the electromagnetic waves will be able to escape the black holes. Therefore, all black holes will gradually lose their status as being black holes, since they will become visible. In other words,

All black holes will gradually become visible.

As the aether pressure in this universe decreases even further, eventually, the pressure of aether in this universe will become equal to the pressure of aether in the accompanying universe. Once this pressure is reached, the flow of aether towards matter particles will slow down to a complete stop. In other words,

"The force of gravity will gradually, literally fade away."

From this instant onwards,

The force of gravity will literally become nonexistent.

This era in the existence of this universe can be referred to as,

"Esmailzadeh Zero Gravity Era"

Once this zero gravity era is reached, all of the matter particles that were trapped inside black holes will be able to freely literally float away, either towards this universe or towards the accompanying universe. In other words,

"All black holes will eventually, literally faint, and let go of all that they have ever collected."

Locating the Birthplace
of the Universe

Locating the Birthplace of the Universe

Man has always tried to understand how this universe has come into existence and how it has evolved. Particularly over the last 200 years or so, using huge telescopes, he has managed to collect and catalogue quite a comprehensive amount of data about thousands of galaxies. The universe is estimated to be about 13.7 billion years old, and it is estimated to be over 93 billion light years in diameter.

According to the observations performed by such scientists as Edwin Hubble, all galaxies are getting away from each other. The speed at which galaxies are receding is directly proportional to their distances. This means that, the farther two galaxies are from each other, the faster they are getting away from one another. The expansion of the universe as a whole, which is still observable and measureable, is the greatest indication that the whole universe had started its existence from a smaller region. It must have expanded in all directions, an expansion process that is still in progress.

By reversing the expansion process of the universe and following the motion of the galaxies backwards in time, as they approach each other, one can visualize the contents of the universe being squeezed into a much, much smaller region, with a specific set of spatial co-ordinates. In other words, this expansion process which is still in progress, had to have taken place at some specific location in this vast expanse of space.

Using the physical evidences made available through the use of telescopes, radio telescopes, X-ray and Gamma-ray telescopes, one can calculate the spatial coordinates of where this expansion process started, billions of years ago.

Two independent methods of performing this very task are introduced in this section. They will lead to identifying the physical location (universal galactic co-ordinates of) where this expansion process started from. These two methods will perform their tasks by using the already collected and available data regarding individual galaxies, on the macro scale. These data include galaxies' locations, apparent shapes as well as their types. Since these two methods are independent of each other, they can also be used to confirm each other's results.

It should be mentioned that, knowing the birthplace of this universe, and even the direction pointing towards that particular location will be of great value to astrophysicists.

1- By knowing the direction which points directly towards the center, astrophysicists will be able to look directly backwards in time.

2- Also by studying the galaxy formations in other directions, astrophysicists will also be able to confirm the validity of their models regarding the formation/development of galaxies as well as their internal/external dynamics.

Background information used in both methods:

The first thing that catches one's attention about the whole universe, on literally every scale, is that planets, stars and even galaxies are demonstrating two types of motions, namely a rotational and a translational motion.

These two types of motions, namely the rotational motion around their own axis and the translational motion around something else, in nearly one hundred percent of the cases, are in nearly the same plane. Their being in nearly the same plane (in most cases) has not been due to chance.

The origins of all spins associated with any of the universe's contents, on various levels of complexity, have started from the time the simplest of all unions were formed between particles.

The rotational motions are due to the differences between the speeds of the individual particles uniting with each other, as they were running away from the center of expansion. In time, as more and more particles (pieces) were drawn together due to their mutual gravitational attraction towards each other, they made up for the differences in each others' speeds by forming a rotational motion around each other. All planets, stars, solar systems and galaxies clearly demonstrate their rotational and orbital motions for all to observe.

Again, it has to be emphasized that, in nearly one hundred percent of these cases, the two types of motions for each planet, star, or galaxy are in nearly the very same plane.

Consider the sun and the planets in the solar system. Based on the data collected through direct observations, majority of the planets (with the exception of Uranus and Pluto) spin around their own axis in a plane which makes less than 30 degrees with their respective orbital planes around the sun. Their orbital planes are important because they literally pass through the sun. Also, the orbital planes of all planets are offset from the sun's equatorial plane by less than 7.16 degrees. For details please refer to the table below.

Planets in the Solar System	Inclination of Rotational Axis, Degrees	Inclination to ecliptic Plane, (Earth=0) Degrees	Inclination to Sun's Equatorial Plane, Degrees
Mercury	0.00	7.00	3.38
Venus	-2.70	3.39	3.86
Earth	23.45	0.00	7.16
Mars	23.98	1.85	5.65
Jupiter	3.12	1.30	6.09
Saturn	26.73	2.49	5.51
Uranus	97.90	0.77	6.48
Neptune	29.56	1.77	6.43
Pluto	122.00	17.20	

Again, it should be noted that, the rotation of all planets is due to the difference in the speed/momentum of their constituents, namely the particles that were thrown into space by the sun. As these particles kept on orbiting the sun, they either united with other particles or were absorbed by existing planets. Either way, the difference in their linear motions caused their unions to manifest a rotational motion which is literally in a plane that is close to their orbital planes around the sun.

Of course, it should be noted that, in case of the solar system all of the planets are imbedded in a relatively thick pancake like volume of space. In other words, the case of the solar system can be considered as being a semi-two-dimensional scenario. However, in this universe, galaxies are spread in a true three-dimensional volume of space.

As it was mentioned earlier, all galaxies clearly demonstrate these two types of motions, as well. Even though, there are lots of spherical galaxies in the universe, there are billions of galaxies which are clearly flat in their geometries.

The rotational motions of all galaxies particularly those that still demonstrate flat geometry, are due to the gathering of constituents that were moving at different rates of speed, as they were literally getting away from the center of the expansion, namely the birthplace of this universe.

By paying attention to the orientations of the galaxies that are flat in their geometries, one can readily try to narrow down the region of space where the expansion process started, billions of years ago. This task can be performed in two different ways.

First Method:

Using the intersection of the galactic rotational planes

In this method one needs to follow the steps outlined below:

Step one:

He needs to make a simple, small model of a planar galaxy or a disk with two pivotal axes, which are at 90 degrees to each other, as shown below.

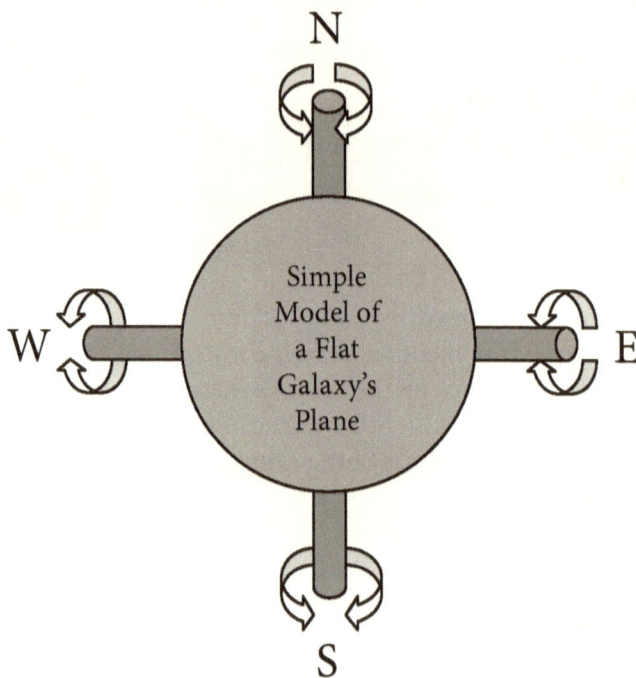

Step two:

He needs to hold this model in such a way that one axis points in the north-south direction and the other in the east-west, as seen directly face-on.

Then, by tilting it on its two axes (one at a time) through different angles, he can duplicate the true angles at which the rotational plane of a given galaxy in mind is rotated, with respect to the line of sight from the earth.

Step three:

He needs to go through all of the available galaxy pictures and sort out the ones which, regardless of their view angles from the earth, still clearly exhibit an overall disk shaped structure.

Then, he needs to use the disk model to duplicate their respective apparent tilt angles about both axes, with respect to earth.

Next, using the degrees at which their respective planes are tilted, he needs to write the equation describing the rotational planes for each and every one of these galaxies, (refer to high school algebra, the section on equations of planes and their perpendicular distances to any given point).

Then, he needs to organize these data in a table and categorize them based on their respective distances from the Milky Way galaxy.

Step four:

He needs to divide the 3-D space of the universe into relatively small blocks or cubes, and designate them with the co-ordinates of their respective center points.

For this purpose he may choose to use any of the universally accepted co-ordinate systems, such as the Galactic or the Super Galactic co-ordinate systems, so long as he stays consistent in all of the co-ordinate data used.

It may be simpler to convert all spherical coordinates into Cartesian coordinates and use straight cubes to define small regions of space.

Step five:

Using the formulas given in high school algebra, one can calculate the shortest distances (perpendiculars) from each and every cube's centers to each and every galactic plane considered.

If the calculated distance is less than one half of the cube's dimensions, it would mean that, if that particular galaxy's rotational plane were to be stretched/extended, it would pass through that particular cube.

This procedure needs to be repeated for every single cube and with respect to every single galactic rotational plane that is considered.

The location of the birthplace of this universe is simply within or quite near the cube(s) which has the most galactic rotational planes passing through it.

These calculations may be performed using a personal computer. The only limitation for any individual is having access to various comprehensive galaxy catalogues.

Note that, to obtain more precision, one needs to use pictures of as many galaxies as possible. The more the number of galaxies considered, will automatically result in a more definite indication of the cube(s) of space which is the closest to the actual location where this universe started its existence.

If need be, one can start with fairly large size cubes to limit the calculation time required. Then, he can use finer size cubes (step by step) to obtain as accurate a result as desired.

Second Method:
Using edge-on view of the galactic rotational planes

In this method, one only needs to pay attention to the apparent orientation of the rotational planes of the galaxies.

The expansion of this universe is a three-dimensional expansion. As particles that were formed were moving away from the center, they possessed different speeds. Therefore, as they were pulled together and formed giant clouds, the difference in their speeds gave rise to the formation of a rotational motion or spin action in these clouds.

Over time, as these giant clouds became more and more concentrated, both in small regions within these clouds and as a whole, the spin was conserved. The smaller localized regions gave birth to stars and eventually planets. And, the overall spin of the giant clouds were conserved, as each one eventually evolved into a galaxy.

The observable rotations of galaxies are clear indication of the spin that existed in the original gas clouds, and also the conservation of this spin, as galaxies have evolved into various shapes and forms.

Therefore, <u>one can expect that most of the galaxies which are flat or planar in their geometries must also have their rotational planes to roughly point towards the center of expansion.</u> That is, if their rotational planes were extended long enough, they would definitely pass through/near the region of space where the expansion had started.

In other words, the rotational planes of most of the flat/planar galaxies which are located along a straight line, between the center of expansion and the outer layer, should be viewed by each other as a narrow strip or edge-on view (or close to it) rather than as a wide plane or full-face view. (Samples are shown below)

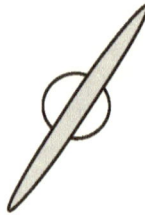   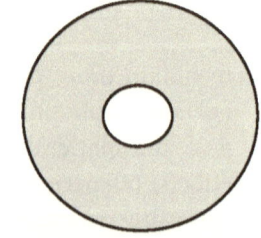

Edge-on view of a galaxy   Full-face view of a Galaxy

Note that, the orientations of these edge-on views are irrelevant. That is, they can be oriented in the north-south, east-west or any other direction, as long as they are viewed as being edge-on or close to it.

Some standard catalogues containing the observational data on galaxies, also include data on the planar tilts of these galaxies, in the form of a ratio of the longer and shorter dimensions of their apparent planes, which is exactly what is needed to perform this task.

As one examines the pictures taken from all of the galaxies which have preserved their planar shapes, he should be able to distinguish two opposing directions in space that would have the highest percentages of the edge-on galaxies in view, as compared to all of the other directions.

Of these two distinct opposing directions, one will be pointing directly towards the birthplace of the universe, while the other will be pointing directly towards the outer layer of the physical universe.

Predictions / Applications

Predictions / Applications

The theories presented in this book provide sound explanations for different basic concepts in physics, all in a consistent way. They are all interconnected and have one thing in common; they are all based on the existence of aether.

They provide certain predictions which enforce their joint validity. They also allow certain applications which otherwise would not have been possible. The following are only a few of such predictions/ applications.

1- Locating the birthplace of the universe
2- Measuring the speed of the aether flow
3- Calculating the pressure difference between the aether in this universe and the aether that is in the accompanying universe
4- The strength of the force of gravity as a function of distance
5- The formation of matter particles gave birth to the force of gravity.
6- The formation of the matter and anti-matter particles caused the slowing down of the initial sudden expansion of this universe.
7- The gravitational force is weakening.
8- The expansion rate of the universe must be gradually decreasing at a rate that is slower than it is expected due to the gravitational forces alone.
9- The expansion rate of the universe seems to be speeding up rather than slowing down.
10- What is the eventual fate/destiny of all black holes?
11- The planetary orbits are widening.
12- All of the solar systems and galaxies will lose their structural integrity.
13- All of the stars will literally turn off and cool down.
14- The force of gravity will be neutralized, altogether.
15- The expansion of aether will never stop.
16- At relativistic speeds, density increases and not mass.
17- The speed of light is affected by the density of the aether.
18- The speed of light (speed of the phase vibrations in aether) is gradually increasing.
19- Time is gradually speeding up in this universe.

20- Time stands still for someone who is travelling at speeds that are equivalent to or faster than the speed of the phase vibrations in aether.

21- Inside a black hole's event horizon, time can be experienced only by objects and beings that are in a free fall directly towards its center.

22- Time cannot be experienced by objects and beings that choose to enter a black hole tangentially, through spiral motion.

23- The frequency of light received from various galaxies is dependent on their positions relative to earth.

24- The frequency dependence of the speed of light in a given matter medium is dependent on the temperature of that medium.

25- Crossflow experiments for two monochromatic light waves or monotonic sound waves

26- Building a <u>time machine to slow down the aging process</u>

27- Building an <u>aether propulsion system</u>

28- Creating bypass routes (wormholes) between specific locations

29- Objects can travel faster than the speed of light.

30- Space is and will be expanding literally forever.

31- Space is finite and it is geometrically spherical.

32- The internal structure of space is like a sponge.

33- All spatial dimensions are straight lines, and are extending as aether is literally continually stretching the boundaries of space.

34- Space, hosts this universe and its accompanying universe.

35- Aether is the only content of this universe.

1- Locating the birthplace of the universe.

This universe is still expanding, an expansion that implies a center. That center can be identified using the already gathered information on galaxies. It has been shown in the section on "Locating the Birthplace of the Universe" that one can use the currently available data to locate where universe started its existence. In fact, two independent methods are presented which can also be used to confirm each others' findings. The two methods are based on:

1- Using the intersection of the galactic rotational planes
2- Using the edge-on view of the galactic rotational planes

For more details please refer to the section "Locating the Birthplace of the Universe"

2- Measuring the speed of the aether flow

This task can be performed using the information provided in different sections of this book.

• The speed of the aether flow induced due to a magnetic field can be calculated using the data collected during the experiment described in the section "What is Magnetic Field".

By measuring the amount of the shift in the sine wave, as seen on the scope of the multichannel analyzer, and knowing the frequency/wavelength of the light wave used, one can calculate how much the phase vibrations of the beam of light were dragged while passing through the apparatus' length.

Knowing the physical length of the magnetized portion of the apparatus (that light passed through), the amount of the signal shift on the scope and the speed of light, one can readily calculate the speed of the aether flow that is induced by the magnetic field.

Also, by repeating the experiment with different voltage settings one can gather data to create a chart that would indicate the exact relationship between the strength of the magnetic field generated and the induced speed in aether.

• The speed of aether induced due to the force of gravity can also be calculated by measuring the amount of deflection of the light beam, as it travels a known distance horizontally (the longer the better). This experiment was described in the section "Gravity and Light".

The very same type of setup can also be used to measure the speed of aether flow that is induced due to the force of gravity of the sun, the moon as well as the center of the galaxy. These experiments have to be carried at different timing and close to the equator, depending on the position of the celestial body of interest relative to earth.

3- Calculating the pressure difference between the aether in this universe and the aether that is in the accompanying universe.

It is known that the speed of the air flowing out of a tank, through an opening such as a valve or a nozzle, is dependent on its pressure as well as the size of the opening. One can readily calculate the required air pressure to promote the speed of the air escaping the tank through a given size of an opening to reach the speed of sound.

One can use the very same methodology to calculate the pressure difference that currently exists between the medium of aether in this universe as compared to the aether that is in the accompanying universe. The needed known facts are as follow:

- Aether has literally zero viscosity.

- The very existence of black holes implies that the pressure difference must be sufficient to promote the speed of the aether flow to reach the speed of light in a vacuum (speed of the phase vibrations in aether) as it reaches the black hole's event horizon, let alone its actual surface.

- The minimum size of the star that can form a black hole is known. This implies that the physical size of the smallest black holes that can possibly form, at the present time, which is also the same as the size of the opening into the accompanying universe is known.

Using the information regarding the minimum possible size of a black hole and knowing that the flow has to reach at least the speed of light as it reaches the black hole, one can readily calculate the minimum pressure difference that must exist between the aether that is in this universe as compared to the aether that is in the accompanying universe.

4- The strength of the force of gravity as a function of distance

The strength of the force of gravity is dependent on the distance to the gravitating object (planet, star and so on). This relationship was given by Mr. Newton and is known as the inverse squared law.

$$\mathbf{F} \approx \mathbf{G}m_1 m_2 / x^{2.000}$$

Where $\mathbf{F}$ is the gravitational force, that two objects of masses m_1 and m_2, located at a distance of x, exert on each other. And, $\mathbf{G}$ is the gravitational constant.

More accurate measurements will indicate that the force of gravity associated with any matter particle or an aggregate of matter particles such as stars and planets (within a galaxy) changes by a bit more than the inverse squared law. This is due to minute reduction in aether density (a gradient in the density profile of aether) as it approaches a large aggregate of matter particles. In other words,

On smaller scales (within galaxies), the force of gravity
changes a little bit faster than the inverse squared law. or,

$$F \approx Gm_1m_2 / x^{2.001}$$

Also, the separation of galaxies is changing due to the expansion of the aether medium (space), as a whole. This effect leads to an apparently weaker force of gravity than expected over long distances. This is due to the fact that, the longer distances are communicated/observed after longer periods of time which include older times, when aether was also denser. In other words,

On large scales (intergalactic), the force of gravity will seem to change slower than the inverse squared law predicts. or,

$$F \approx Gm_1m_2 / x^{1.999}$$

5- The formation of matter and anti-matter particles gave birth to the force of gravity.

The force of gravity, as it is presented and described in the section "What is Gravity?", is the drag force that is exerted by aether as it is flowing towards matter and anti-matter particles. Therefore, the force of gravity has started its existence since the matter and anti-matter particles were formed. In other words,

"Matter particles (in general) literally gave birth to the force of gravity."

Note that, in the beginning, as aether was experiencing its rapid expansion, there was no force of gravity, since matter and anti-matter particles were not formed yet.

6- The formation of the matter and anti-matter particles caused the slowing down of the initial sudden expansion of the universe.

The very formation of matter and anti-matter particles and their sudden abundance to the point of saturating the whole medium of aether caused two major side effects:

1- Due to their abundance, matter and anti-matter particles allowed unrestricted flow of aether from this universe into the accompanying universe. In a relatively short period of time, the aether pressure was reduced drastically, the same pressure that was the driving force for aether's expansion.

2- Due to their abundance, matter and anti-matter particles generated/ imposed such a strong gravitational force towards each other that literally acted like a braking system and slowed their expansion process in the whole universe.

As most of the matter and anti-matter particles united and annihilated each other, leading to severe reduction in the force of gravity present, the leftover matter particles continued with their expansion process in this universe, but at a much slower pace.

In other words,

"The formation of matter particles (in general) literally drastically slowed down the spreading of the contents of the universe, and the expansion of aether, and hence space."

7- The gravitational force is weakening.

The force of gravity which is in fact the drag force induced due to the flow of aether is directly dependent on aether's density, as well as its pressure.

Therefore, as the density/pressure of aether is reduced, due to both expansion and leakage, its flow is becoming less effective in inducing the force of gravity. In other words,

"The so-called universal gravitational constant is actually dependent on aether's density and pressure, and it is decreasing with time."

This gradual reduction in the force of gravity will in turn have certain definite consequences throughout this vast universe.

8- The expansion rate of the universe must be gradually decreasing at a rate that is slower than it is expected due to the gravitational forces alone.

As time goes on and the overall aether density and pressure are reduced everywhere, the drag force exerted by aether flow on any matter particle is also reduced. This means that, the overall force of gravity is gradually becoming less effective. The affecting factors are:

1- The expansion of the universe, both in aether and with it.

- The effect of distance on the strength of the gravitational force is calculated using the formula provided by Mr. Newton, namely the inverse squared law. As two objects get farther away from each other, they automatically exert a weaker force on each other.

2- Aether's overall expansion, as well as its leakage out of this universe is reducing its pressure and its density in this universe.

- The effects due to decreases in aether's pressure and density were not accounted for in Mr. Newton's formula. Therefore, these two additional factors must be taken into account, because they directly affect the effectiveness of the force of gravity in slowing the expansion rate of this universe. Note that,

    — **The lower aether pressure** simply means less push or driving force to encourage the flow of aether as it is escaping through any available opening. The less aether escaping through the opening automatically <u>translates into lower aether speeds approaching the opening</u> from different directions, hence weaker drag force is induced as other particles are encountered.
    — **The lower aether density** <u>also lead to less effective drag force induced by the flow of aether</u>, as it encounters other particles in its path.

    This automatically means that the expansion process of this universe is actually slowing down at a slower pace than would be expected/predicted by any gravity theory that is not time dependent and therefore does not take the drop in the density and pressure of aether, and their direct effects on the force of gravity, into account.

9- The expansion rate of the universe seems to be speeding up rather than slowing down.

The expansion process of the universe, which is based on direct observations, is graphically presented below. According to these findings, the universe seems to be accelerating in its expansion process, rather than decelerating as it is expected, based on the known laws of gravity.

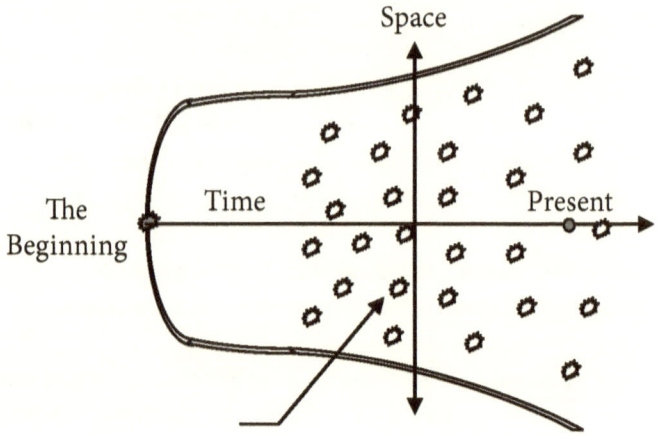

As it was explained in (# 8) above, the overall force of gravity is gradually becoming less effective due to two different reasons:

1- The expansion of this universe which is literally increasing the separation of the galaxies from each other. The farther the galaxies get from each other, the weaker they affect each other gravitationally.

2- The reduction in aether density and its pressure due to both its overall expansion as well as its leakage out of this universe.

These factors directly influence the effectiveness of the force of gravity in slowing down the overall expansion rate of this universe.

Also, when considering long distances, such as billions of light years, due to longer time spans involved, some other factors also come into play. These factors are:

1- The gradual increase in the speed of light (speed of the phase vibrations in aether), which is due to aether density gradually decreasing,

2- The gradual increase in the pace at which time is experienced, which is due to increase in the speed of the phase vibrations in the aether medium while the relative motions of celestial bodies are not changing as much.

When all of these effects are taken into account, the observed behavior can be readily explained.

Consider two galaxies, one located at about one billion light years and the other about one half of one billion light years away from earth. They can be detected to be receding from earth at say about 1,000,000 km/hr and 600,000 km/hr, respectively. The observed discrepancy is not due to the universe accelerating in its expansion, but rather due to the numeric value of the speed detected for the farther away galaxy. That galaxy's speed should be corrected for the variations that have occurred during the past one billion years.

Once the proper corrections have been implemented, it will become clear that the speed of the farther away galaxy is actually much higher than it seems to be at the present. In other words,

It is due to the increase in the speed of light and the rate at which 'time' is experienced, as well as the gradual reduction in the density and pressure of aether, that lead to such an illusion that, farther away galaxies are not receding as fast as they should be. Therefore, their lower speeds than expected, give the impression that the universe as a whole is speeding up in its expansion, while it is actually slowing than, but at a slower pace than expected when static gravitational theories are used.

10- What is the eventual fate/destiny of all black holes?

All black holes will eventually cease to be black holes due to a natural phenomenon. That natural phenomenon is the overall pressure difference that exists between the aether in this universe and the aether in the accompanying universe.

As time goes on, the aether pressure in this universe is lowered due to both its expansion in this universe and its leakage through matter particles and black holes. In the mean time, the aether pressure in the accompanying universe is continuously raised due to receiving aether from this universe. Therefore, the difference between aether pressures in these two universes is gradually decreasing.

At a certain point in the expansion process of this universe, the difference between the aether pressure in this universe and the aether pressure in the accompanying universe will approach a critical value. That value of the pressure difference corresponds to the minimum required pressure that would cause the flow of aether through the largest possible black holes to reach the speed of light (the speed of phase vibrations in aether). This critical aether pressure difference can be referred to as:

"Esmailzadeh Critical Aether Pressure Difference"

Note that, three factors affect how fast/soon this critical aether pressure difference will be reached:

1- The rate at which the overall pressure of aether in this universe is reduced, as aether is expanding and leaking out,

2- The rate at which the overall pressure of aether in the accompanying universe is raised, as aether is received from this universe, and

3- The rate at which the speed of the phase vibrations in aether is increased, due to aether density decreasing in this universe.

As the Esmailzadeh critical aether pressure difference is approached, all of the black holes, starting with the smallest ones, will lose their status as a black hole, since the local aether pressure difference will no longer be sufficient to cause the speed of aether to reach the speed of light by the time it reaches black holes' surfaces.

Once this critical value of the aether pressure difference has been reached, all of the black holes, regardless of their sizes, will cease to act as black holes, since the speed of the incoming aether reaching their surfaces will be just under the speed of light. This in turn means that, none of the black holes will possess an event horizon, and light and other electromagnetic waves will be able to propagate against the incoming flow of aether and escape. In other words,

Eventually, all black holes will become visible.

As the aether pressure in this universe decreases even further, eventually, the pressure of aether in this universe will become equal to the pressure of aether in the accompanying universe. Once this pressure is reached, the flow of aether towards matter particles will slow down to a complete stop. In other words,

"The force of gravity will gradually, literally fade away."

From this instant onwards,

The force of gravity will literally become nonexistent.

This era in the existence of this universe can be referred to as,

"Esmailzadeh Zero Gravity Era"

Once this zero gravity era is reached, all of the matter particles that were trapped inside black holes will be able to freely literally float away, either towards this universe or towards the accompanying universe. In other words,

"All black holes will eventually, literally faint, and let go of all that they have ever collected."

11- The planetary orbits are widening.

As the aether's density is decreasing over time, due to its expansion as well as leakage (to the accompanying universe), it is becoming less effective in dragging objects that happen to be in its path. Therefore, stars are gradually, exerting less of a gravitational force on their respective planets. This is in turn leading to the widening of the orbital path of planets around stars.

One can say that, stars are gradually losing their grip on their planets and consequently

"The planetary orbits are gradually becoming wider."

This side effect of decrease in aether's density has already been detected (in regards to the planet earth). According to the collected data, the orbital path of the planet earth is widening by about 7 meters (about 23 feet) per century.

Of course, a small portion of the observed amount is justifiably due to the sun losing mass as it is continuously transforming some matter into energy and also as it is literally throwing some matter into space as solar storms.

Note that, the widening of the orbital path of planets around their respective stars will lead to eventual and definite **GLOBAL COOLING** on all planets. This global cooling, which will actually take place on all planets existing in this universe, can be referred to as,

"Scharback Universal Planetary Cooling"

12- All of the solar systems and galaxies will lose their structural integrity.

As the pressure/density of aether in this universe is reduced further and the force of gravity becomes even weaker, stars will no longer be able to hold on to their respective planets. Stars also will no longer be motivated to stay in their own orbits around the centers of their respective galaxies, either. Therefore, solar systems and galaxies will lose their structural integrities and all planets and stars will relatively freely float in space. In other words,

**"All of the celestial bodies will become
independent wanderers in this universe."**

13- All of the stars will literally turn off and cool down.

As the force of gravity becomes weaker, the outer layers of stars will no longer weigh as much. Therefore, they will not exert enough pressure on the inner layers to promote the ongoing nuclear fusion reactions. At this point, stars will literally turn off and start to cool down. They would basically become cold gas giants.

As all of the stars, one by one, turn off, **the lights in the night sky will gradually become less in numbers and eventually all will disappear, forever.** Once all of the stars are turned off, the whole universe will start its **eternal dark era.** In other words,

"The universe will become pitch black"

Also, as stars such as our sun turn off, **their respective planets like our earth (if they are still in orbit) will start experiencing a night that is eternally long.**

14- The force of gravity will be neutralized, altogether.

As the pressure/density of the aether in this universe is reduced further and the force of gravity becomes even weaker, all of the celestial bodies will lose their structural integrities. At that point, all of the planets and stars in the universe will fall apart.

Eventually, the pressure of the aether in this universe will equalize the pressure of the aether in the accompanying universe. Once this pressure is reached, the flow of aether towards matter particles will slow down to a complete stop. In other words, as the pressure difference between this universe and the accompanying universe approaches zero,

"The force of gravity will gradually, literally fade away."

From this instant onwards,

The force of gravity will literally become nonexistent.

At that point in time, if the universe has any momentum left in its expansion, it will keep on expanding forever, since it would literally lack any kind of braking instincts. This era in the existence of this universe can be referred to as,

"Esmailzadeh Zero Gravity Era"

From this point in time onwards, none of the existing matter particles will exert any gravitational force on each other. In fact, even the matter particles that were trapped inside black holes will be able to freely literally float away, either towards this universe or the accompanying universe.

This final stage will eventually lead to a totally homogenous distribution of matter particles everywhere, as they will be freely floating in total harmony and peace, without exerting any kind of force towards each other, since all subatomic particles with opposite charges will unite and neutralize each others' charges.

15- The expansion of aether will never stop.

The volume of the aether medium will always keep on expanding, even after its pressure in this universe and in the accompanying universe have equalized. This expansion process will literally continue until the universal density/pressure of aether reaches zero. This density/pressure will be realized only when the size of the universe reaches infinity.

16- At relativistic speeds, density increases and not mass.

As an object's speed approaches the speed of light in a vacuum, it contracts in the direction of its motion. However, the other two dimensions, that are perpendicular to its direction of motion, remain unchanged. Therefore, as the speed of an object approaches that of light (speed of the phase vibrations in aether), its volume decreases linearly with its length, that is in the direction of motion. This automatically implies that, not the mass, but the density of the object increases as its length shrinks, due to relativistic effects. See figures below.

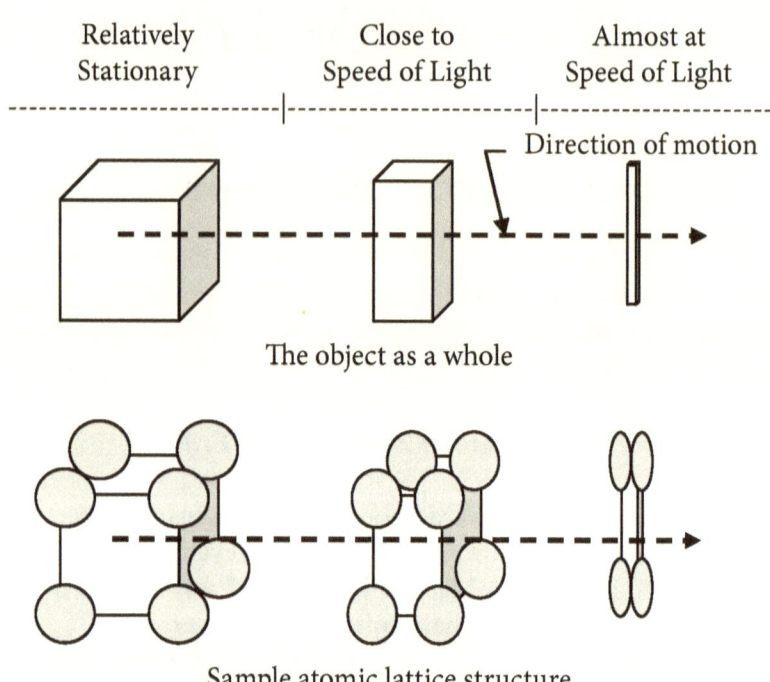

The object as a whole

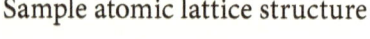

Sample atomic lattice structure

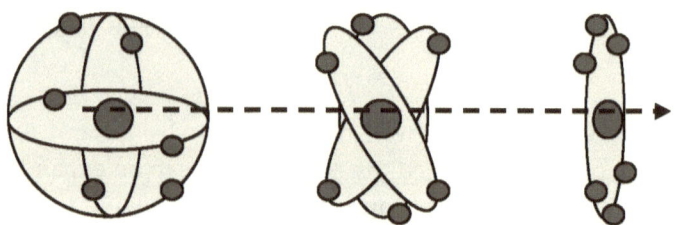

The individual atoms

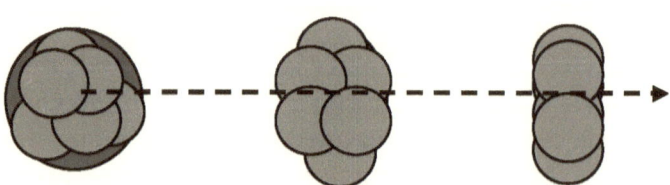

The individual nuclei and their nucleons

Note that, the individual subatomic particles do not get squeezed in the direction of motion.

In other words,

"As an object speeds up and approaches the speed of light in a vacuum, its density approaches infinity and not its mass. Its mass remains unchanged."

- The energy that is utilized to increase the object's speed relative to its local aether is in fact stored in the object in different forms. The kinetic energy of the object that is increased due to its speed, accounts for some of the energy consumed. However, most of the energy is stored as different potential energies that are shared between various components of the object, as they are literally compressed towards each other and flattened out of their normal shapes. Just like the energy that is stored in a spring as it is compressed, or as the air inside a cylinder is squeezed in one particular direction, by pushing on the piston.

Note that, as an object's speed approaches that of the phase vibrations in the local aether, even though its mass stays unchanged, but its weight increases, just as if it is experiencing a stronger gravitational field. This is due to the fact that, in both cases, the aether is rushing by at higher speeds. And,

"If the object manages to reach the speed of light in a vacuum, it will literally experience having an infinite weight. This is equivalent to standing still at the event horizon of a black hole, where the speed of aether rushing by is exactly equivalent to that of the phase vibrations in that medium."

The above figures clearly demonstrate the variety of deformations that are induced in the internal structure of an object, as its speed relative to its local aether approaches the speed of the phase vibrations in that medium. These deformations are literally due to the push/drag that is exerted by the aether as it is made to rush by, either because the object is moving in aether or aether is encouraged to move relative to the object.

Note that, if an object stands relatively stationary and aether is encouraged to pass by it, the effect will be just the same. For example, if an object is placed in a strong gravitational field, a strong magnetic field or a strong electric field, it will experience the very same effects on its density as if it were travelling at high rates of speeds in aether. In other words,

"The relativistic effects on density and weight of an object depend on that object's motion relative to its local aether medium."

17- The speed of light is affected by the density of the aether.

The speed of light and other types of phase vibrations in aether are dependent on aether's density. Higher aether density means that the phase vibrations can travel slower in that medium.

18- The speed of light (speed of the phase vibrations in aether) is gradually increasing.

Light and other electromagnetic waves are phase vibrations in aether and are affected by the density of that medium.

Therefore, as the density/pressure of aether is reduced, due to both the overall expansion of aether and leakage of aether through matter particles, the speed of the phase vibrations in that medium is continuously increasing. In other words,

"As the density/pressure of aether is decreasing the speed of light and other phase vibrations in aether are increasing."

In other words,

"The speed of light is gradually increasing."

The consequences of this effect are quite important in two regards, namely the distant past and the distant future.

- **In the distant past, the speed of light was quite low.**
  To calculate what the speed of light was back in the beginning, one needs to know what the density of aether was back then. The absolute value of aether's density cannot be calculated. However, its value can be estimated relative to what it is at the present time.
  Currently the universe is estimated to be about 93 billion light years across. If, in the beginning, the size of the universe was ONE

meter in diameter, then the density of aether in the beginning can be estimated as follows,

At the present time the diameter of the universe, in millimeters is, $(8.80 \times 10^{26} \, m)$, since,

$(9.3 \times 10^{10}$ light years$) \times (365$ days/year$) \times (24$ hours/day$) \times (3600$ seconds/hour$) \times (300,000$ km/sec$) \times (1,000$ m/km$) = 8.80 \times 10^{26} \, m$

Therefore, the volume of the universe has grown by the cube of that number which is (6.81×10^{80}). This number also indicates the ratio of the density of the aether medium at the beginning relative to what it is at the present time.

Note that, this number (6.81×10^{80}) is an estimated value, due to the following:

1- Some aether has escaped this universe, over its life.

2- The initial size of the universe as being one meter is only a guessed number.

3- The current size of the universe is an estimated value.

Since, in the beginning, the density of the aether medium was about (6.81×10^{80}) times higher than it is at the present time, and also knowing that the speed of light and other phase vibrations in aether are dependent on the density of aether, one can confidently make the following statement,

"The speed of light during the very first 'second' that this universe started its existence was literally billions of billions of times slower than what it is at the present time."

- **In the distant future, the speed of light will approach infinity.** Since, as time goes on, the density of aether is continuously decreasing and it is headed towards zero. This in itself implies that the speed of light and other phase vibrations in that medium are correspondingly approaching infinity. In other words,

"The speed of light and other phase vibrations in aether will ultimately approach infinity."

19- Time is gradually speeding up in this universe.

The rate at which the passage of time is experienced by any object is dependent on the speed of that object relative to the local aether. As this relative speed approaches that of the phase vibrations in aether, the time is experienced at a slower pace. However, if this relative speed becomes a smaller fraction of the speed of the phase vibrations in aether, the time experienced will be at a faster pace.

As the aether is expanding, its density is gradually decreasing. This reduction in aether's density leads to higher phase vibration speeds in that medium. Also, the speeds of objects (galaxies) relative to aether are decreasing, since the overall expansion process of this universe is slowing down with time due to the force of gravity.

In other words, gradually, the speed of objects relative to aether is becoming a lesser and lesser fraction of the speed of phase vibrations in it. Therefore, the overall effect is that,

"The rate at which the passage of time is experienced
in this universe is gradually speeding up."

In other words,

"Time is speeding up."

The consequences of this effect are quite important in two regards, namely the distant past and the distant future.

- **In the distant past, time was progressing quite slowly.**
  To calculate how slow time was actually progressing back in the beginning, one needs to know what the density of aether was back then. The absolute value of aether's density cannot be calculated. However, its value can be estimated relative to what it is at the present time.

  As it was shown in the previous section, the density of aether in the beginning was about (6.81×10^{80}) times higher than it is at the present time. According to that ratio, one can readily estimate that the rate at which 'time' was experienced back then must have been literally billions of billions of times slower as compared to its current rate. Therefore, one can confidently make the following statement,

"The very first 'second' that this universe started its

existence, was literally as long as millions if not billions
of years long, if measure using today's time scale."

- **In the distant future, time will progress at an incredibly high pace.**
Since, as the density of aether is continuously decreasing and it is
headed towards zero, the speed of phase vibrations in that medium
will correspondingly approach infinity. This automatically means
that the speed of objects will become smaller and smaller fractions
of the speed of phase vibrations in aether. Therefore, the passage of
time will be experienced at a faster pace.

  Eventually, as the speed of phase vibrations in aether approaches
infinity, the rate at which time is experienced by all objects in this
universe will also approach infinity. In other words,

  **"Gradually the duration of minutes, hours, days, years,
even millions of years will become equivalent to the
duration of a 'second' as it is experienced today."**

20- Time stands still for someone who is travelling at speeds that are equivalent to or faster than the speed of the phase vibrations in aether.

Anyone who manages to reach the speed of phase vibrations in aether, or
even be able to go any faster, will only experience the present. He/she cannot
skip into future that has not arrived yet.

Notes:
- The 'Future' can be experienced only when it becomes 'Present'.

- The only effect on the occupants of a spaceship capable of speed
of light (or even beyond that) is literally limited in slowing down
and even stopping the aging process, but the 'time' experienced will
always be 'Now'.

21- Inside a black hole's event horizon, time can be experienced only by objects and beings that are in a free fall directly towards its center.

This is due to the fact that, they will be moving at that speed with aether
and not in it. In fact, their motion relative to aether will be minimal. Therefore,
they do experience time just as fast as individuals who are watching them fall

into the black hole. However, the duration of such an experience will be next to nothing due to their high speeds and the shortness of the distance to the surface of the black hole. The figure below demonstrates such a scenario.

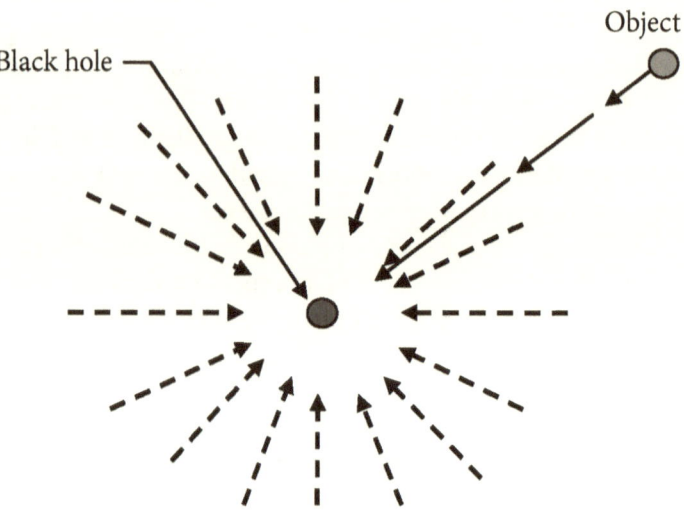

Arrows indicate aether flow from all directions

22- Time cannot be experienced by objects and beings that choose to enter a black hole tangentially, through spiral motion.

Time will slow down until it stops altogether, for such items, before they even reach the event horizon, since they be moving relative to aether (almost at 90 degrees) and not with it. The figure below demonstrates such a scenario.

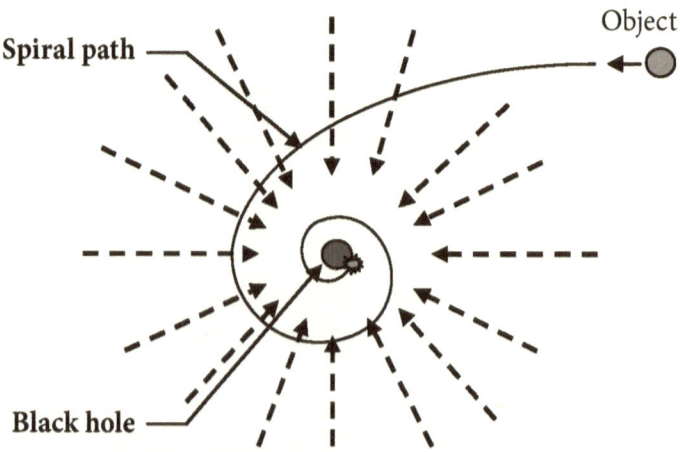

Arrows indicate aether flow from all directions

23-The frequency of light received from various galaxies is dependent on their positions relative to earth.

This is due to the fact that, the density of aether that is passed through along the way is different, in different directions.

24-The frequency dependence of the speed of light in a given matter medium is dependent on the temperature of that medium.

This is due to the fact that, as the temperature of a certain matter is changed, its atoms and molecules vibrate at a different frequency. The other effect caused due to the change in temperature is the changes that occur in the spacing between individual atoms and/or molecules in the matter, as the matter expands/contracts.

For example, as the temperature of a cube of copper, iron and so on is raised, their atoms and molecules vibrate at faster rates and their overall spacing also increases.

25-Crossflow experiments for two monochromatic light waves or monotonic sound waves

Consider two beams of light of the very same frequency directed so that they cross each others' path. By adjusting the timing (shifting) of the wave cycle of one wave as compared to the other (by changing the length of the wires, using a sliding contact), one can obtain a variety of unexpected colors.

The very same type of experiments can also be conducted with monotonic sound waves.

26-Building a <u>time machine to slow down the aging process</u>

Applying the information provided in this section and the effects demonstrated in the section "What is Magnetic Field?", one can build a sort of time machine which in fact would slow down the time experienced by any object (including living beings) that happen to be placed inside it.

To accomplish this task, there is no need to build a starship capable of flying at near the speed of light to go for an interstellar journey and/or orbit a black hole and risk being pulled in by its force of gravity. One can accomplish this intended task of experiencing the time dilation right here on earth.

To do so, he/she needs to create a strong and constant magnetic field in a closed space such as a room (which can serve as the bedroom, office, living room or even the whole house) where the objects of interest including living beings can spend most of their times in and so benefit from its generated effects. The following drawing shows a simple presentation of such a device.

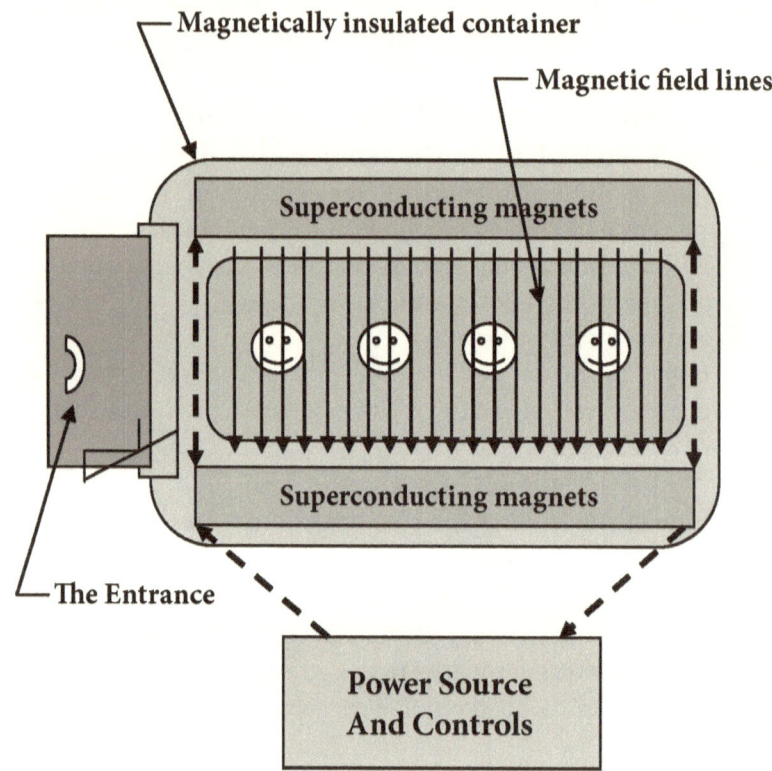

Note that, in this case, the objects or living beings do not have to be locked in that specially equipped limited space. They can come and go as they please. The longer time an individual spends in the space equipped with special effects, the more of an age difference he/she will experience as compared to those who do not enter such a device/space.

For more details in this regard, please refer to the book "Innovative Inventions" by the author of this book.

27- Building an aether propulsion system

The main component in this type of propulsion system is basically one (or more) superconducting magnet(s) that is used to direct the flow of aether in a specified direction, literally like a jet engine.

The overall effectiveness of this type of propulsion system depends on the strength of the magnetic field generated. Since the effects generated by such a system are accumulative, increasing the length of the magnetized region will automatically increase the overall speed of the aether flowing through. This is exactly like stacking up regular inline water pumps to increase the output pressure.

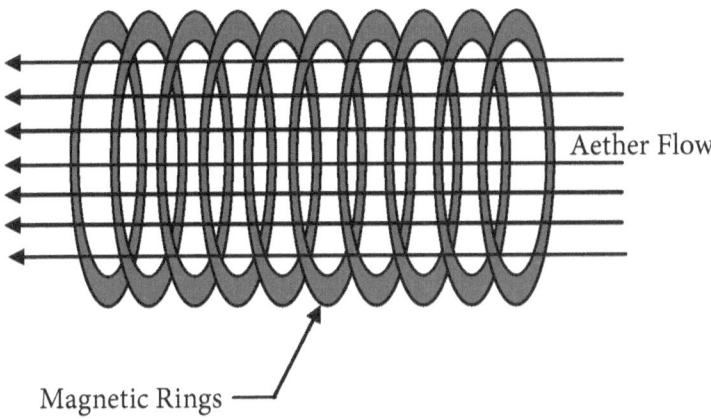

Aether Flow

Magnetic Rings

For more details in this regard, please refer to the book "Innovative Inventions" by the author of this book.

28- Creating bypass routes (wormholes) between specific locations

By manipulating either the electric field, magnetic field or a combination of the two, bypass routes (wormholes) can be made to form between specific locations. In other words, matching portholes can be formed to exchange objects through the accompanying universe. This exchange process will take place at speeds that can be readily mistaken for instantaneous.

The key operations/processes that will enable one to succeed in the manufacturing of such devices will be "Quenching/Implosion" (as well as "Resonance/Harmonics") of the magnetic and/or electric fields so that temporarily, just for the required duration, the speed of aether flow reaches the speed at which the phase vibrations travel in the local aether medium.

**"Sudden extreme variations in the electric current
at highest voltages possible can allow such needed
quenching/implosion of the magnetic field."**

To boost a device's capabilities, one can use the accumulative effect (resonance/harmonics) of electromagnetic field generated.

29- Objects can travel faster than the speed of light.

Light and other electromagnetic waves are phase vibrations in aether, just as sound is a form of phase vibration in a medium such as air. Therefore, as it is stated in the section on light, **speeds attainable by objects are not limited to the speed of light in a vacuum** (vacuum being the aether medium that is void of any matter particles).

This is exactly like objects' speeds not being limited to the speed of sound in the medium of air.

Therefore, even though light cannot get out of a black hole, due to its being literally carried by the aether flow, **an object can fight its way out of a black hole**, given it has the proper structural composition and some kind of aether propulsion system.

Other types of propulsion systems will not be able to function properly in that environment and will not be as effective, either. This is due to the fact that, the propulsion mechanism used must be able to literally manipulate the aether environment that it is in.

30- Space is and will be expanding literally forever.

In the beginning, space was contained in a very small volume, because aether was in its super compressed state. Due to some phase vibration that was introduced into it, aether started a sudden expansion. As aether expanded, so did the vastness of the space which was occupied by aether. Over time, the space became the size it is observed to be, today.

Space (and not just the physical universe) is still expanding, and will keep on expanding forever. This is due to the fact that, aether itself is expanding. However, aether's expansion rate, which is directly related to its pressure, is decreasing with time.

Note that, the outer boundaries of space, or the front line of the aether's expansion is even farther ahead of the phase vibrations (electromagnetic waves such as light) that are spreading in that medium.

Even though, due to the force of gravity, the expansion process of the matter contents of this universe may gradually slow down to a halt at some point in the future, but the expanding aether will keep on spreading the frontiers of space.

The expanding rate of aether will gradually reduce as its pressure in this universe is reduced. But, its expansion process will never come to a complete stop, since the pressure of aether will only theoretically approach zero. In other words,

"The radius of the space that is hosting this universe is and will be expanding at ever slower rates and yet it will never come to a complete halt, ever."

31- Space is finite and it is geometrically spherical.

The overall geometrical shape of space is spherical, since aether's expansion was symmetric in all spatial directions. Aether's overall front line is a sphere which is expanding like a balloon. This automatically indicates that the overall size of space is finite. One may ask:

<u>What is aether expanding in? Isn't that the true space?</u>

These questions are more philosophical questions than they are related to the physical universe. They are just like the questions that one can ask regarding the source of aether and what was the source of that source, and so on.

Any answers to such questions can only be based on pure speculation. This is due to the simple fact that, in order to even know what the outside of this universe (or even a house, for that matter) looks like, let alone where it is located, one needs to literally see it from the outside. Because, even if the whole of the interior of a house (for example) can be surveyed and precisely mapped out, still due to the lack of knowledge about the thickness of different exterior walls, one would not be able to even guess what the outside of the house may look like.

32- The internal structure of space is like a sponge.

The internal structure of space can be analyzed on two different scales:

- **On the microscopic scale,**
  On the microscopic scale, the internal structure of space is quite like a sponge which has countless number of tiny holes in it. These holes are in fact the matter particles that have formed due to phase vibrations in aether. These holes literally act as drain holes through which aether is escaping from this universe into some other dimension, the accompanying universe.

- **On the macroscopic scale,**

   On the macroscopic scale, space is divided into two distinct regions. One region encompasses this universe, and the other encompasses the accompanying universe. These two regions are connected through countless number of microscopic passage ways through which aether is allowed to flow from this universe which is at a higher pressure into the other one that is at a lower pressure.

33- All spatial dimensions are straight lines, and are extending as aether is literally continually stretching the boundaries of space.

The topographical presentations of the gravitational forces near massive stars and even galaxies give the impression that the spatial dimensions are curved, while they are not. The curvature that is detected is not due to the spatial dimensions themselves but rather due to the gradients that exist in the flow speeds of the local aether that is headed towards the local star, the galaxy or even a black hole.

Stars and galaxies (on their own scales) are dense matter concentrations in relatively small or limited regions of space. Denser matter concentrations mean more available drain holes for aether to flow through, in a given volume of space. This effect automatically translates into higher flow rates of aether towards that region of space. As it is explained in details in the section on gravity, the speed of the aether flow is also dependent on the distance to the matter particles (or their aggregates, such as stars or even galaxies).

Particularly, the drawings representing the flow speeds of aether in the vicinity of black holes clearly demonstrate a funnel shape with an increasing slope at smaller distances from the black holes' event horizons.

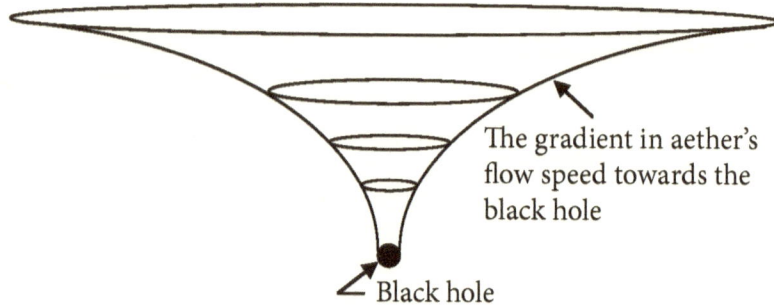

The gradient in aether's flow speed towards the black hole

Black hole

The flow of aether is from all directions

However, the gradual increase in the slope of the funnel shaped surface towards its center, shown in the figure, does not represent any kind of curvature in the spatial dimensions, but rather the gradient that exists in the ever increasing speed of the aether as it approaches the black hole's event horizon.

The very same is also true about the gravitational influences of other celestial bodies such as stars and galaxies. However, in their cases, the central portion does not become so pronounced in its slope.

34-Space hosts this universe and its accompanying universe.

Aether, which actually defines space, is a continuous medium that not only occupies this universe, but also the accompanying universe.

The two universes are connected through the matter particles which act as drain holes for aether to flow from this universe into the other one.

35-Aether is the only content of this universe.

Aether was the first content of this universe. Over time, and due to different circumstances, aether manifests its many faces which include various forms of matter and energy.

For instance, based on the information presented in this book, it is proposed that matter (including anti-matter) is only a condensed form of aether, as it experiences spikes in its fabric. Also, the electromagnetic waves in general are phase vibrations in aether. In other words,

"Everything in this universe including matter, anti-matter, dark matter, as well as all forms of energy, including dark energy and vacuum energy, are just different manifestations of aether, the one and only content of this universe."

Conclusions

According to the theories presented in this book, aether exists. It is an environment in which all of the contents of this universe are literally floating. Aether is also responsible for carrying the variety of forces that are exchanged between matter particles.

It is clearly demonstrated that aether is not a stationary medium as it was assumed by the nineteenth century physicists. It is quite a dynamic medium. The force of gravity as well as the forces associated with the magnetic and the electric fields encourage the formation of a variety of flow patterns in aether, from the most microscopic scales to the most macroscopic scales.

The flow patterns in aether are just like the variety of flow patterns that are experienced by the molecules of air in the atmosphere or by the molecules of water in the oceans.

In the beginning, aether was the only content of this universe. Over time, and due to different circumstances, aether manifests in many different states which include various forms of matter and energy. In other words,

"Everything in this universe including matter, anti-matter, dark matter, as well as all forms of energy, including dark energy and vacuum energy, are different manifestations of aether, the one and only fundamental content of the physical universe."

Or,

"Aether is the only content of this universe."

And, finally,

"Once the existence of aether is accepted and its effects are taken into account in explaining various forces that exist in this universe, one can readily arrive at the well overdue, and very much sought after **GRAND UNIFIED THEORY.**"

References

Books:

1- J. C. Maxwell, "A Dynamical Theory of the Electromagnetic Field", Philosophical Magazine, (1864).
2- A. A. Michelson and E. M. Morley, Philosophical Magazine, (1887).
3- A. Einstein, "Relativity, the Special & the General Theory", Translated by R. W. Lawson, Methuen & Co. Ltd. London, (1920).
4- D. C. Giancoli, "Physics", Prentice Hall, Inc. Englewood, Cliffs, New Jersey, (1985).
5- E. T. Whittaker, "Theories of Aether and Electricity", Hudges, Figgis & Co. Ltd,, Dublin, (1910).
6- B. Esmailzadeh, "Innovative Inventions", (2012).
7- B. Esmailzadeh, "Innovative Theories", (2011).

Internet:

1- http://en.wikipedia.org/wiki/Galaxy
2- http://en.wikipedia.org/wiki/Kepler%27s_laws_of_planetary_motion
3- http://en.wikipedia.org/wiki/Newton%27s_law_of_universal_gravitation
4- http://www.jrank.org/space/pages/2433/Lick-galaxy-catalogue.html
5- http://heasarc.nasa.gov/W3Browse/all/rc3.html
6- http://en.wikipedia.org/wiki/Fritz_Zwicky
7- http://en.wikipedia.org/wiki/Dark_matter
8- http://en.wikipedia.org/wiki/Dark_energy
9- http://metaresearch.org/cosmology/cosmology.asp
10- http://nedwww.ipac.caltech.edu/level5/Shapley_Ames/frames.html
11- http://en.wikipedia.org/wiki/Age_of_the_universe
12- http://en.wikipedia.org/wiki/Observable_universe
13- http://skyserver.sdss.org/dr1/en/astro/universe/universe.asp
14- http://www.phys.unsw.edu.au/einsteinlight/jw/module3_M&M.htm
15- http://en.wikipedia.org/wiki/Lorentz_transformation
16- http://en.wikipedia.org/wiki/Special_relativity
17- http://en.wikipedia.org/wiki/Black_hole
18- http://en.wikipedia.org/wiki/Inclination
19- http://nssdc.gsfc.nasa.gov/planetary/factsheet/

www.ingramcontent.com/pod-product-compliance
Lightning Source LLC
Chambersburg PA
CBHW031838170526
45157CB00001B/348